W9-BDV-615

SCIENCE AND SOCIETY

# Mother Nature's Pharmacy

## POTENT MEDICINES FROM PLANTS

J. S. Kidd and Renee A. Kidd



Facts On File, Inc.

This one is for Elliana.

---

**Mother Nature's Pharmacy: Potent Medicines from Plants**



Facts On File, Inc.
11 Penn Plaza
New York NY 10001

**Library of Congress Cataloging-in-Publication Data**

Kidd, J. S. (Jerry S.)
Mother Nature's pharmacy: potent medicines from plants / J. S. Kidd and Renee A. Kidd.
p. cm.
Includes bibliographical references and index.
Summary: Discusses the uses of plants as medicines and the role of government in protecting public health and doing research for new medical treatments.
ISBN 0-8160-3584-9
1. Materia medica, Vegetable—Juvenile literature. 2. Medicinal plants—Juvenile literature. 3. Pharmacognosy—Juvenile literature. [1. Pharmacology. 2. Medicinal plants.] I. Kidd, Renee A. II. Title.
RS164.K53 1998
615'.32—dc21 97-37925



Text and cover design by Cathy Rincon
Layout by Robert Yaffe
Illustrations on pages 18, 42, 57, 79, 92, and 102 by Jeremy Eagle

Printed in the United States of America

MP FOF 10 9 8 7 6 5 4 3 2

This book is printed on acid-free paper.

# Contents

# Acknowledgments

We thank the administrators, faculty, and staff of two organizations devoted to higher education—The College of Library and Information Services of the University of Maryland, College Park, and the Maryland College of Art and Design. In particular, Dean Ann Prentice and Associate Dean Diane Barlow at College Park have been extraordinarily patient and supportive.

We are also grateful for support and guidance from colleagues at the National Academy of Sciences/National Research Council in Washington, D.C. Again, special thanks to Anne Mavor and Alexandria Wigdor for their kindly dispositions and to Susan McCutchen for her high spirits.

Lastly, Professor Richard Evans Schultes gave us a gentle nudge in the right direction at the right time.

# Introduction

This book is one in a series about the relationship between science and society. Many features of that relationship are revealed by examining how plants are used as sources of medicine. The science of botany is central to the topic, but other areas of science such as anthropology and chemistry are also involved. Society plays a role in the relationship through a variety of organizations and institutions. However, the role of the central government in protecting the public health and conducting the search for new medical treatments is emphasized.

Today, almost everyone is concerned about health care and the medicines taken to preserve or restore good health. Therefore, each person needs a basic understanding of all types of medicines—new and old—and how they affect a person's well-being. Without this information, it is difficult to ask relevant questions and to make good decisions about health care.

The search for plant-derived medicines has gone on for many centuries. Over the years, scientists have tested hundreds of plants from the earth's temperate zones. However, few plants native to tropical rain forests have received the same attention. The current search for health-giving plants is concentrated on the thousands of varieties that thrive in tropical regions. Indeed, botanists have discovered many promising plants in the rain forests. The discovery of new medicinal plants has important consequences for patients, for the native peoples of the region, and for the conservation of rain forests.

Many new medicines are discovered and developed for use every year. The pace is so rapid that it is difficult to keep up with changing methods of treatment. Society's acceptance of certain medicinal substances also changes. For example, cocaine, produced from the leaves of the coca tree, is now a controlled substance. This means that the legitimate use of cocaine is limited to licensed health-care providers such as doctors. This was not always the case. In the early 1900s, cocaine was available both as a prescription medicine and as an ingredient in popular beverages. These drinks were sold without restrictions and consumed daily by many people to improve their health. Now, the beverages would be illegal because we know that cocaine is a powerful, harmful, and addictive drug.

A drug can be defined as a medicine used in the treatment of disease or it can be defined as a controlled substance because of its toxic effects. Many of the controlled substances are narcotic. A *narcotic*, from the Greek word for numbness, is any drug that dulls the senses, produces sleep, and causes the user to become addicted to the substance. Some narcotics used as sleeping pills can be helpful but many have few, if any, positive benefits.

All drugs—whether medicines or controlled substances—can cause harm as well as good. During the time of the Renaissance, Paracelsus, a Swiss physician, advocated the use of plants as sources of medicine. About 1500, he recognized that all medicinal products—including those derived from plants—could be dangerous. Paracelsus realized that the dosage of medicine must be carefully monitored. He wrote: "It is dose alone that makes a poison."

This statement is true of some commonly used substances. For example, coffee is so widely consumed that people tend to forget that it contains a mild drug called caffeine. Caffeine is an alkaloid—a complicated carbon-based molecule that affects the nervous system. If taken in concentrated form, caffeine produces symptoms of heart disease. Even consumed as a far less concentrated beverage, too many cups in one day can have a noticeable effect on the nervous system. For some people, coffee is mildly addictive, and headaches or other discomforts can appear when its consumption is halted.

The use of alcoholic beverages provides a dramatic example of narcotic effects. Alcohol is a habit-forming depressant. In small quantities, it lessens tensions and promotes digestion of food. However, if the amount is even modestly increased, alcohol can impair the functioning of the central nervous system. The drinker finds it difficult—if not impossible—to drive a car safely or to operate machinery in the proper manner.

Societies often decide which substances are regarded as medicinal. A substance considered healthful in one culture might be defined as a poison in another. For example, few Europeans or Americans have been persuaded to consume a sushi made from raw puffer fish, known as *fugu*. (Sushi is a Japanese dish that often contains sliced raw fish.) When made with puffer fish, it is regarded by some as a great delicacy that has an added thrill. If not properly prepared, the fish can be deadly to eat. In Japan, many males overlook that danger and view fugu as a type of medicine that will increase their sexual potency.

The subject of medicine is of great interest to the general public and of vital importance to the national economy. In the United States, the production and sale of medicines accounts for about $44 billion every year. Experts estimate that about 25 percent of that amount is linked to medicines made from plants—the subject of this book.

In the United States, the pharmaceutical industry plays a major role in discovering, studying, manufacturing, and marketing medicines. However, other organizations are also prominent in the search for new medicines. Medical schools employ scientists who conduct basic research on new compounds and new approaches to disease treatments. Research centers and universities that study pharmacology, chemistry, molecular biology, and microbiology play a crucial role. In addition, many government agencies are involved in developing and testing medicines.

One of the most influential governmental agencies is the gigantic Department of Health and Human Services. This department operates the National Institutes of Health (NIH).

Within NIH are facilities where scientists study the effects of drugs on the body. Institutes affiliated with NIH sponsor many of the drug-related research projects that are conducted in various universities.

The Food and Drug Administration (FDA) is also a part of the Department of Health and Human Services. The FDA oversees all testing and evaluation of new (and sometimes old) medicinal compounds to determine if each substance is safe and effective.

Other U.S. government agencies that influence the development of new medical drugs include the Environmental Protection Agency (EPA), the National Science Foundation, and the Small Business Administration (a part of the Department of Commerce). Departments of the military services that deal with health care and chemical warfare are also involved with medicinal products. State and local governments are concerned with the certification and licensing of pharmacists and with the regulation of medicines used at health-care facilities such as nursing homes.

Fortunately, consumers have some determination in how the products of the huge pharmaceutical industry are made available and utilized for medical treatments. Informal influences are exercised by the general public, health-care professionals, and the mass media. Individuals involved in special interest groups and institutions such as medical schools and organized religion also exert influence. However, governmental agencies at the local, state, and federal levels have far greater authority over the delivery of pharmaceutical products.

All of these individuals and groups are important to this account of plants used as medicines. However, botanists and biochemists are the most significant people in this story. Botanists travel the world in search of new medicinal plants, and biochemists transform these plants into medicines to prolong human life.

1

# Progress in Health Care

Long before historic times, humans used natural products as medicines. In all likelihood, the use of plants for medicinal purposes grew out of the search for plants that were good to eat. This quest might explain the use of garlic as a medicine. Anyone who has ever attempted to eat raw garlic remembers its strong taste and smell. Ancient cooks, too, might have noted that it was not a pleasing food when eaten by itself. Eventually, they realized that the plant was flavorful when cooked with other ingredients. Early medicine men and women, however, may not have regarded garlic as a possible food. These healers reasoned that raw garlic's strong smell might make it a powerful medicine against illness. Even today, some people consider garlic a medicine that helps relieve various health problems.

For more than 50,000 years, humans have experimented with eating many things—both plant and animal. If the experiment revealed a bad-tasting, sickening, or deadly substance, the information was passed from one generation to the next. In the same way, knowledge about substances that helped or cured sick people was communicated to others. Treatments using folk remedies were passed on to successive generations long before the invention of written records.

In some human societies, the results of successful medical treatments were not widely shared. Instead, secrecy was prevalent. Shamans were often considered to have special gifts or powers. Some groups believed only those with these special spiritual connections could have certain knowledge. Because of this shamans sometimes maintained a monopoly on information about medicinal products. Because of their knowledge, they enjoyed a privileged position within their group. Although most shamans were well-intentioned, some took unfair advantage of those who were ill, refusing to use their learning unless they received favors from patients or the families of the sick.

Today, such knowledge is no longer kept secret. The ethics of science requires that accounts of all scientific investigations must indicate the research methods, materials, findings, and conclusions. Therefore, anyone can duplicate the work and arrive at the same result. When applying for a patent, the situation is similar. The inventor must reveal the procedures necessary to accomplish the invention. The inventor's rights are protected by the patent, but the details of the assembly of the device or the steps in a process are made public.

## Early Medical Practices

For thousands of years, most medical treatments relied on the use of plants. The ancient founders of science-based medicine were herbalists—people who studied and used plants—who recorded their knowledge in scrolls and books. Greek pioneers such as Hippocrates (c. 460–c. 377 B.C.) and Galen (A.D. 129–c. 199) combined the skills of botanist and pharmacist with those of diagnostician and physician.

In general, ancient physicians were gentle when caring for the sick. They kept the patient warm and dry, and often treated him or her with medicines made from plants. One version of the Hippocratic oath, still recited by newly licensed medical practitioners, contains the phrase, "above all, do no harm." However,

*Alchemists and apothecaries were principal sources of drugs in the Middle Ages.* (Courtesy of the Photographic Archive Services of the National Library of Medicine)

during the Middle Ages (period in Europe from c. 500 to 1500), the European approach to medical treatment changed and many physicians adopted a "heroic" response to disease. These treatments were often far more intrusive and painful.

To some extent, heroic methods were based on the ancient theory of humors. This theory was advanced in the time before the Christian era and was still in vogue in the early 1500s. According to this theory, the human body contains four key fluids, or humors, which must be kept in balance to maintain good health. The humors were blood, phlegm, black bile, and yellow bile. A personality trait was associated with each of the humors. Someone who was relatively cheerful and active was designated as sanguine. Sanguine people were thought to have a surplus of blood. In contrast, a phlegmatic person was depressed and sluggish. These conditions indicated an excess of phlegm. People now know that these notions about humors are not true.

However, to the practitioners of times past, these ideas provided direction for diagnosis and treatment. For example, if someone became overly active and agitated during an illness, the person was diagnosed as sanguine. Since this indicated too much blood, bloodsucking leeches were placed on the patient's body to remove some of the blood. Other treatments to bring the humors into balance included induced vomiting—by mechanical or chemical means—and deep irrigation of the large bowel. These heroic responses were unpleasant, unsanitary, and often fatal.

Other tactics to cure sickness were based on restoring a balance between the conditions of temperature and dampness. To relieve various ills, patients were sometimes wrapped in very cold, wet cloths and sometimes in hot, dry ones. This method was uncomfortable—but much less injurious than those used to keep the four humors in balance. The health benefits thought to come from steam baths and saunas may have been derived from this theory.

Other heroic treatments were developed during the 19th century. After electricity became available, some patients were treated with electroshock to various parts of the body. The location of the shock depended on the disease and was intended to improve the physical or mental health of the sick person. Other patients were forced to ingest disagreeable substances. In those days, some caregivers reasoned that a painful or disgusting treatment would force the patient to recover to avoid further treatment.

## Reforms

In most cases, treatments based on physical abuse were not effective. Consequently, compassionate physicians sought alternative treatments. In the 1780s, William Withering, a Scottish physician, noticed that country folk in England made a herbal tea from the leaves of the foxglove plant. The tea was given as a treatment to those who suffered from abnormal swellings and heart pains. Withering decided to study the use of foxglove leaves as a medicinal substance. His experiments proved that foxglove did help heart conditions, and Withering developed the heart medicine now known as digitalis.

To study the properties of his new medicine, Withering tested digitalis on 163 patients with heart disease. He was one of the first doctors to undertake a systematic evaluation of a therapeutic substance. After Withering's pioneering work, the use of clinical trials for testing new medicines became a common practice. This system led eventually to the modern techniques of drug testing.

*Digitalis from the foxglove plant is an effective treatment for congestive heart disease. It was identified by exploring folk medicine.* (Courtesy of the Photographic Archive Services of the National Library of Medicine)

Humane physicians used other means to oppose cruel heroic methods. In the 1790s, Samuel Hahnemann, a respected teacher of medicine, advocated bed rest and nourishing food to cure illness. He believed that the human body had extensive powers of recovery and should be allowed to heal itself whenever it could.

Some farsighted individuals investigated methods to prevent people from becoming ill. In the mid-1700s, Lady Mary Montagu, an English poet, toured the eastern Mediterranean. She returned to England with news of a new technique to protect humanity from smallpox—a very contagious, dangerous, and disfiguring disease. While traveling in Turkey, Lady Montagu learned about the technique of inoculation from local Muslim physicians. Their method was to give the patient a mild form of the disease and hope that the person could recover. Once cured, the patient would never be reinfected with the disease. However, this method was dangerous. Some of those inoculated developed a full-fledged case of smallpox and died. The technique of inoculation needed to be perfected before it could be used safely.

In the last quarter of the 18th century, William Jenner, an English country doctor, discovered a safe method to inoculate people against smallpox. Jenner had many dairy workers among his patients. Over the years, he observed that cowpox, a common disease among cattle, had some interesting properties. Jenner noted that dairy workers did not became very ill when they caught cowpox from their cows. After they recovered from cowpox, the same workers rarely caught smallpox. The physician reasoned that a case of cowpox somehow protected the person against smallpox. Around 1780, he injected healthy people with a serum containing a substance taken from the sores of sick cows. The patients developed cowpox but later proved to be immune to smallpox. Although no one—including Jenner —understood why this technique was effective, it became widely used. Over the years, Jenner's discovery saved thousands upon thousands of lives.

*William Jenner grasped the idea that immunity from a particular disease developed by surviving an infection of that disease. He did so without knowing the cause of infections.* (Courtesy of the Photographic Archive Services of the National Library of Medicine)

## Lifestyles

In the United States during the 1830s, several charismatic leaders promoted an unusual idea of disease prevention. Each promised that a radical change in lifestyle would result in a long, disease-free lifetime. Sylvester Graham, the most popular—and one of the most extreme—of these leaders, was a traveling preacher.

Graham laid down specific rules to attain good health. He advocated that all his followers adhere to a vegetarian diet that included a nourishing whole grain biscuit that he had devised—the graham cracker. In addition, his rules included drastic restrictions on the relationship between men and women.

*Louis Pasteur discovered that microscopic creatures could cause disease.* (Courtesy of the Photographic Archive Services of the National Library of Medicine)

When Graham died at the relatively young age of 57, his bold system of life reform met with some skepticism. However, other more realistic reformers followed Graham. Moderation was the approach advocated by leaders such as W. K. Kellogg in the early 1900s.

## The Application of Science

By the mid-1800s, advances of a more scientific nature—especially in the field of biology—led to a better understanding of disease. In 1858, the German biologist Rudolph Virchow used a microscope to demonstrate that the progress of a disease could be observed in the affected cells. During the 1850s and 1860s, the French scientist Louis Pasteur showed that microorganisms—organisms that can be observed only under a microscope—were responsible for many diseases. His research proved that a specific disease was caused by a specific microorganism. Pasteur reasoned that if the disease-causing microbe could be killed, the disease would be cured.

The German scientist Robert Koch, too, pursued the idea that a particular microbe was the cause of a particular disease. In 1882, he determined the identity of the tuberculosis microbe that attacks the cells of the lungs. In 1883, he identified the cholera microbe that attacks cells in the digestive system. These discoveries motivated chemists such as the German bacteriologist Paul Ehrlich to spend a lifetime searching for specific remedies. Around 1900, after many false starts, Ehrlich formulated a compound of arsenic called Salvarsan 606. This medicine was lethal to the microbe that causes syphilis—a dangerous, sexually transmitted disease that attacks cells in the nervous system.

Although new medicines and techniques were in limited use by the turn of the century, doctors, government officials, educators, and the public were dissatisfied with the available medical care. By that time, most people realized that neither lifestyle reforms nor heroic techniques were producing good results.

*Paul Ehrlich was noted for his persistence. He tried hundreds of compounds before discovering an effective treatment for syphilis.* (Courtesy of the Photographic Archive Services of the National Library of Medicine)

Many poorly informed physicians still resorted to painful treatments. In the early 1900s, partly in response to the general dissatisfaction, a new, scientific approach to health care became popular. This approach gained increasing acceptance after scientists achieved control over yellow fever, botulism, and other deadly diseases. These and other important discoveries radically changed the practice of medicine in industrialized countries.

## Science and Medical Education

Medical schools were the first institutions to make major changes. Before 1900, much of medical training was an apprentice system in which physicians allowed students to observe and copy their procedures. After 1910, medical schools in the United States required course work in the basic sciences in addition to clinical observation. Soon, the whole concept of medical education began to change. At first, medical students were required to complete three years of premedical college work. As medicine became a more prestigious profession and the number of applicants increased, medical school administrators sought to reduce the number of would-be physicians. They increased the required period of undergraduate study to four years. These years laid a solid foundation in chemistry and biology in addition to courses in physics and mathematics.

After students were admitted to medical school, the first two years were spent in advanced science training. This training included courses in physiology, human anatomy, and other areas that are necessary to an understanding of the human body. The last two years focused on an extensive study of specific diseases.

In the early days of medical education reform, aspiring doctors rarely saw patients for diagnosis or treatment. In many cases, seven years of preliminary study were required before a young doctor had much contact with a patient. In more recent times, starting in the 1970s, health-care managers complained that new doctors were qualified scientists but knew little about the real needs of patients. Gradually over the past 25 years, many medical schools have begun to involve students with patients throughout their years of medical education prior to their internship experiences.

Today, other changes are being introduced in medical education. For many years, most medical students became general practitioners after graduation. These men and women took care of all medical needs including delivering babies, setting bones, and curing diseases. By the 1950s, physicians began to specialize in one area such as heart disease or intestinal problems. At present, U.S. government funding for medical education and health care is supporting the reduction of narrow specialization. Government officials believe that doctors associated with general or family practices can provide better basic health care.

The early 20th-century reforms have had many consequences. The long years of medical training and the high cost of medical education has led to a gradual increase in the income and social status of physicians. The required courses in science have assured better educated and more competent doctors. The reforms also helped to make the profession more effective and efficient. Many dreaded infectious diseases—such as scarlet fever, mumps, and whooping cough—have become far less of a health hazard. New techniques and medicines have reduced deaths due to diseases such as AIDS and cancer—and may soon develop cures for these illnesses. Vaccines and other protective measures have helped to control communicable diseases such as influenza and measles. Today, computers link doctors and other health-care providers so that they can obtain worldwide medical information on patients, medicines, and procedures.

In the last 50 years, nontraditional medicines and medical techniques from non-Western cultures have found a place in Western health care. Indeed, the National Institutes of Health have investigated the clinical value of Chinese acupuncture—a method once scorned in the West. In this ancient medical treatment, trained technicians insert thin, sharp needles in specific areas of the body to relieve the effects of certain ills. Many patients believe this procedure improves their health.

2

# The Early Plant Hunters

People have been hunting for and using plants for thousands and thousands of years. Indeed, there are early records of medicinal plants and the remedies made from them. An image on the wall of an early Egyptian tomb shows men presenting a variety of exotic plants to a pharaoh. A clay tablet from ancient Mesopotamia, known to be more than 4,000 years old, contains the oldest record of medical prescriptions. An Egyptian papyrus, written about 500 years later, provides recipes for medicinal compounds made from plants. Ancient Chinese scholars, too, contributed to the store of knowledge. A Chinese emperor, in the third century B.C., discovered many new medicines—some of which are still in use.

The ancient Greeks acquired their medical information from these earlier civilizations and added new techniques and medicines of their own. The Romans accepted and further expanded this learning. Throughout the Greek and Roman times, the ancient tradition of collecting and identifying plants was continued. Medicinal plants were identified and listed in scrolls and, later, in books for the use of scholars and medical practitioners.

After the fall of Rome in A.D. 476, however, learning and experimentation were all but forgotten. The works of philosophers, poets, and scientists were in danger of destruction from neglect. Waves of invaders from the border lands between

Europe and Asia flowed into western Europe. Governmental and other institutions collapsed or were abandoned. Precious writings were saved by two groups of people. In the West, many monks and nuns who served the Christian church spent their lives copying and preserving the manuscripts. In the Muslim empire—especially along the northern coast of Africa and in Spain—the ancient learning was transcribed and protected by Islamic, Jewish, and Christian scholars. During the Middle Ages (c. 500 to c. 1500), medical scholars from the Muslim-dominated areas greatly increased the store of medical knowledge—especially in the fields of chemistry and medicinal products.

After 711, when Islamic armies invaded and conquered Spain, learned Muslim Arabs, Jews, and Christians joined together to forge a golden age in Spain. These scholars—aided by the religious tolerance prevalent at that time and place—were responsible for great advances in the arts and sciences. Indeed, during those times, Spanish scientists had the most advanced knowledge of medicine and medical botany. This golden age began to wane in the 1100s when the Muslims turned against the Jews. At that time, Jewish scholars and physicians began to leave Spain and migrate to other countries. Later, when Muslim dominance was threatened by armies from central Europe and local Spanish knights, further migrations of learned scientists took place. Southern Italy was an attractive refuge. The new migrants made the Italian peninsula a stronghold of medical education and practice. As early as the 11th century, the Italian city of Salerno had became the home of the first full-fledged medical school in Europe. According to legend, the school was founded by four masters—a Western Christian, an Eastern Christian, a Jew, and a Muslim.

During the High Renaissance (in Italy, early to mid-1500s), a period of great intellectual productivity, botanical interest in foreign plants was centered on ornamentals, plants cultivated for beauty. By the 1500s, northern Europe was prosperous and people wanted showy gardens. New medicinal plants were still of interest, but most people believed that their apothecaries—who bought and sold herbs—knew everything about curing patients.

## Rauwolf, Plant Hunter

Some scientists, however, continued to seek out and investigate new medicinal plants. One such scientist was a German botanist and physician, Leonhard Rauwolf. Because of his medical training, he was able to document his findings in a scientific and systematic manner. Rauwolf was born in Batavia about 1535. He did his preliminary studies at the University of Wittenberg, a school not far from Berlin. In 1560, he began his advanced medical training at the University of Montpelier in southern France. Montpelier was then the finest medical school in Europe, and he received a splendid education.

Rauwolf was born during the Renaissance—a historical period that lasted from the 1300s until the end of the 1500s. During the Renaissance—a word that means rebirth—Europeans rediscovered the classical knowledge of the Greeks and Romans. Rauwolf was required to study the works of ancient authors such as Socrates, Plato, and Aristotle. The works of these and other Greek writers had long ago been translated into Latin and Arabic. The ancient Greek and Roman works generated great interest in many areas of human concern—including the sciences. Scientific writings and the works of the early physicians, such as Hippocrates, Dioscorides (a master herbalist), and Galen (a renowned anatomist and herbalist) were in demand.

The works of these physicians were studied by all medical students. In keeping with the ancient medical theories, standard treatments were almost exclusively herbal—all medicinal substances were made from plants. However, diet restrictions, purging of the bowels, and bleeding by either mechanical means or by leeches were a part of many treatment programs.

European students of medicine in Rauwolf's day were well versed in the Bible as well as knowledgeable about classical learning. During their Bible study, they noted the mention of each medicinal plant or herbal treatment. Because of this, a typical student had a special interest in medical practices that had originated in the Holy Lands of the eastern Mediterranean area. This area includes Greece, Turkey, and the biblical lands that are now called Iraq, Syria, Lebanon, Saudi Arabia, Jordan, and Israel. In Rauwolf's time, the biblical lands were all ruled by the Ottoman emperor of Turkey.

Rauwolf, like his fellow students, easily combined his deep interest in religion with his scientific studies. Most students believed that herbal medicines and their beneficial effects were all part of God's plan. During his studies, Rauwolf developed a strong desire to visit the Holy Lands. He wanted to explore those areas of the Ottoman Empire that would confirm the written descriptions of biblical plants and their surroundings. Rauwolf also looked forward to a pilgrimage that would deepen his religious faith.

Rauwolf's plans could not be fulfilled for several years. He knew that he must complete his medical studies before he could explore foreign lands. While applying himself to this task, he spent his free time investigating the botanical specimens found in the area of rural France near his medical school. Rauwolf's field trips netted a collection of 443 specimens, which he dried and mounted on white paper. Today, this collection can be seen at the museum of the University of Leyden (Leiden) in Holland.

In 1563, Rauwolf completed his medical education and looked forward to opening a medical practice in Augsburg, Germany. Two years later, he married Regina Jung, the daughter of a prominent Augsburg physician. In the next few years, Rauwolf twice moved his practice to other towns. In 1570, he and his family returned to Augsburg, and the city officials gave Rauwolf a yearly stipend to serve as a public health adviser. During these years, Rauwolf retained his interest in botany. He cultivated both rare plants and the medicinal herbs with which to treat his patients.

For a long time, Rauwolf had desired to observe and collect plant specimens in their native surroundings. In 1572, when he was in his 30s, he decided to travel to the Holy Lands. To fulfill this ambition, however, he needed to find a source of financial support. Rauwolf's brother-in-law owned a trading company that did business in the Near East and was able to supply that support. He employed Rauwolf to act as physician to the company's representatives in that area. The brother-in-law hoped that Rauwolf could keep the employees in good health and perhaps, discover plants that might prove valuable to the trading company. At that time, the idea of generating profit as a direct result of scientific study was an unusual concept. Historians believe that Rauwolf's arrangement might have been the first of its kind.

In the 1500s, France under King Francis I was the only European country to have a trade agreement with Suleiman, the Ottoman Turkish ruler. By the provisions of the treaty, France was allowed to extend its trading rights with the Ottoman Empire to any interested European citizen. The Germans who hired Rauwolf thus did business in the Mideast under the protection of the French.

On September 2, 1573, Rauwolf sailed from Marseilles, France, for the city of Tripoli in present-day Lebanon. After more than three long weeks at sea, the travelers arrived near Tripoli.

The city of Tripoli, built on the slopes of Mount Lebanon, is some distance from the port. By the time Rauwolf and the others reached the city gates, the night curfew was in effect and they were arrested. They were accused of trying to set the city afire with their lanterns. Fortunately, the French consul was passing the gate and the German traders were released into his custody.

At first, Rauwolf and his companions found Tripoli to be a cramped and unattractive city. However, they soon found that the outside of the houses—with their windowless walls and small doorways—were very different from the inside. The walls often hid expansive gardens, beautiful patios with fountains, and attractive inner rooms.

The city was a busy transit point for goods sent from distant places such as India and China. Tripoli was also a major silk production center. Years before, large colonies of silkworms had been imported from China. Mulberry bushes, also imported from China, were grown locally to feed the silkworms.

After about five weeks in Tripoli, Rauwolf was ready to move on. He had enjoyed seeing the various ornamental plants and learning about mulberry culture. Now, he wanted to begin his serious study of plant life and visit the Holy Lands. Company business dictated that the German traders next stop would be the city of Aleppo. This ancient city, some miles west of the Euphrates River, is in northern Syria near the modern border with Turkey. Aleppo had long been a trading center with commerce in spices, cloth, precious stones, gold, and silver.

Rauwolf started his exploration of the area as quickly as possible, and two important plants were brought to his attention. He immediately began an investigation of Chinese sarsa-

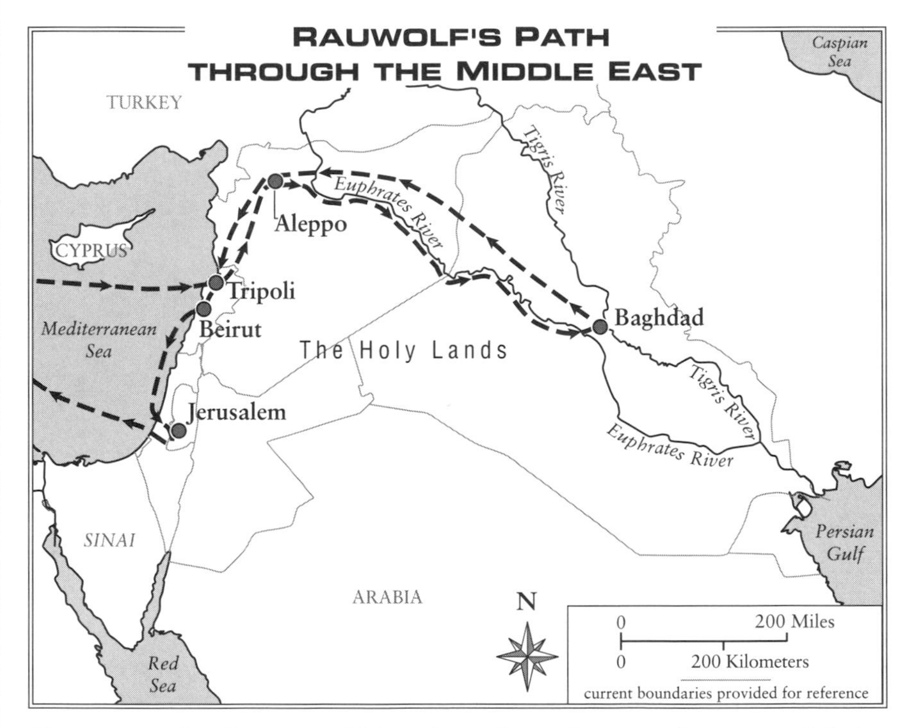


*The routes taken by Rauwolf from Tripoli to Aleppo and on to Baghdad and back are shown on this map.*

parilla and Iranian rhubarb. He learned that the sarsaparilla plant helps cure venereal disease and that the roots of Iranian rhubarb are a remedy for many intestinal disorders and liver complaints. In addition to medicinal plants, Rauwolf studied other useful plants. He was the first European to describe the preparation of coffee and the Arabs' use of coffee on social occasions.

While in the area of Aleppo, Rauwolf collected many plants that were foreign to Europe. He remained there for nine months and then began the dangerous journey to Baghdad in present-day Iraq. Rauwolf and his companions traveled east from Aleppo to the headwaters of the Euphrates River. Next, they sailed down the Euphrates in a southeasterly direction toward the place nearest to Baghdad. That city lay about 100 land miles (160 km) east of this point on the western bank of the Tigris River.

Each evening on their voyage down the Euphrates, the crew pulled into shore and made camp. Even if he was tired, Rauwolf explored every campsite for new plants. However, good finds were scarce in the arid climate along the northern reaches of the Euphrates River. Some of his best finds were interesting species of gourds. When the gourds were dried and pulverized, the powder provided material for a strong laxative.

For the last leg of the journey to Baghdad, they left the relative ease of river travel. Much of the 100-mile (160-km) trip between the two rivers was through stark desert. Deserts are often difficult to travel through, but Rauwolf's main worry was a possible encounter with Bedouin tribespeople who were unhappy with the rule of the Turkish sultan.

The group eventually arrived safely at their destination on the Tigris River. They found the area around Baghdad to be quite fertile because it received irrigation from the river. In this well-cultivated area, Rauwolf found few plants that were unknown to the Europeans. After a short stay, the party prepared to return to Aleppo by a land route. Rauwolf and his companions joined a caravan and traveled by horseback in a northerly

*Plant names in past centuries were not always consistent. What the Europeans called nardus is probably the same plant mentioned in the Bible as spikenard. This plant was among those sought by Rauwolf in the Holy Lands.* (Courtesy of the Photographic Archive Services of the National Library of Medicine)

direction for hundreds of miles up the east bank of the Tigris River. They then proceeded westward until they reached Aleppo.

The travelers received bad news when they reached Aleppo. Rauwolf's brother-in-law's trading company was bankrupt and the company's representatives in Tripoli were being held in prison. Although the rest of his party continued on, Rauwolf decided to remain for a time in Aleppo. To support himself, he became the family physician to the many resident European traders.

Rauwolf's final adventure in the Ottoman Empire was a trip to Jerusalem, the heart of the Holy Lands. At that time, wars and political unrest had left the city in poor condition. Rauwolf was disappointed with this situation and remained in Jerusalem for a very short time. While there, he continued his collecting and found a few unusual plants in the area.

It was now time to return to Augsburg. Rauwolf experienced several further adventures during his journey home. On his sea voyage, he successfully avoided pirates and bad weather. During high winter, Rauwolf safely crossed the dangerous Brenner Pass in the Swiss Alps. He arrived in Augsburg on February 12, 1576, two years and six months after his departure from Marseilles.

All in all, Rauwolf's ventures were successful. He benefited from the fame brought by the publication of his memoirs. His collection of 364 preserved plant species from the Middle East was a major contribution to the botanical studies of the time. Consequently, Rauwolf gained a high reputation within the community of botanical scholars. Indeed, by the time he died in 1596, the story of his life had reached legendary status.

Rauwolf's story did not end with his death. More than 100 years later, in the early 1700s, Europeans were thoroughly exploring the Americas. Plant hunters in the Caribbean area discovered many slightly different varieties of known types of plants. However, some of the specimens appeared to belong to a previously unknown group. The German botanists who found the new type decided to name it in honor of Rauwolf. The newly discovered type of plants became the genus *Rauwolfia*.

Over the next 150 years, the story of *Rauwolfia* became increasingly complicated. Botanists uncovered many new plants within the new genus named after Rauwolf. Members of this genus—a related group of plants—grow wild in all tropical or semitropical zones. In fact, members of the genus were seen to prosper under a wide variety of soil and moisture conditions.

## Reserpine

In India, one species of *Rauwolfia* had been used for centuries to treat snakebite. This use probably originated because the root of the plant resembles a snake. Native healers often linked the treatment of a disease to a medicinal plant that somehow resembled the cause of the illness. In this case, the snakebitten patient was given the ground or powdered snake-shaped root of the *Rauwolfia* plant.

Another traditional Indian use of powdered *Rauwolfia* root was to ease mental disorders. Probably this use was discovered accidentally when a person with mental problems was given a dose of the root for a snakebite. The attending healer might have noted its effectiveness in calming the patient. European physicians in the 1700s, however, were unimpressed with any substance that they regarded as a folk remedy. They did not try this treatment to calm mentally ill patients.

Although *Rauwolfia* was long recognized in the official Indian list of drugs, it received no Western-style scientific investigation until the 1930s. Then, Indian physicians and biochemists, trained in European methods, began to take an interest in the plant. The results of their controlled studies showed convincingly that the *Rauwolfia* powder was an effective tranquilizer. Their findings gave European scientists new incentives for further investigations. In 1952, a German chemist isolated the active ingredient, an alkaloid he called reserpine. With the pure drug to study, it soon became evident that small doses could effect a profound change in the behavior of agitated psychiatric

Rauwolfia, *or Indian snakeroot, has been used in folk medicine as a treatment for snakebite, but a far more important and effective product of the root is reserpine, a natural tranquilizer.* (Courtesy of the New York Botanical Garden)

patients. An hour or so after treatment, the patients became calm and untroubled. This condition allowed patients to respond to other forms of psychotherapy including psychoanalysis. Equally important, it was found that negative side effects were few if the drug was administered properly. Reserpine was seen to be a natural tranquilizer.

The main source of the drug is *Rauwolfia serpentina*, a short, shrublike plant that belongs to the same family as the periwinkle —also known for its medicinal properties. This member of the genus *Rauwolfia* has become far more important than any of the nearly 400 plants that Rauwolf found on his trip to the lands of the Bible.

Its importance in the treatment of mental illness can only be understood when compared to treatment that it replaced. It is particularly useful in the treatment of agitated depression where treatments once included electroshock therapy and convulsive therapy induced by the administration of large doses of insulin. Now, almost 50 years after the adoption of reserpine in conventional medical practices, there are many alternative tranquilizers —some more effective than reserpine. Today, mentally disturbed people often receive reserpine in combination with other medicines. This treatment seeks to promote the total well-being of the patient.

3

# The Framework of Folk Medicine

In industrial societies, most people visit a trained caregiver when they need medical treatment. Usually, the caregiver has been qualified for that position under strict regulations. These health-care providers have completed a lengthy, prescribed program of education and training in their chosen fields. They have been trained as physical therapists, nurse practitioners, cancer specialists, family physicians, or in any one of a long list of specialties. At the completion of their programs, they are tested on their knowledge and understanding of the principles of health care. After passing rigorous tests, each care provider receives a license to enter into professional practice.

The provider may establish a solo practice and offer health-care services as an independent professional. Increasingly, however, qualified people accept positions in a setting where care is provided by a team of professionals who represent a wide range of specialties.

Not all people seek medical help from a professional caregiver. Indeed, people may decide to practice self-care for a variety of health considerations. One example is the person who decides to stop smoking or chewing tobacco. Other modes of self-care typically involve dieting or following a program of physical exercises. However, the most common self-care practice is the use of nonprescription medicines such as aspirin, antacid remedies, and lotions for treating muscle aches or skin conditions. Many herbal remedies, sold as dietary supplements rather than medicines, are also self-prescribed to restore or improve health.

Folk medicine offers several alternatives to those who do not choose to seek a licensed caregiver. Some people may decide to prepare their own medicines from herbs. Recipes for the preparation of such medicines can be found in books available at public libraries and bookstores. Medicinal plants grow wild in woods or fields, and they can be planted in home gardens. However, descriptions of the wild plants are often somewhat vague. This leads to the unfortunate prospect that the untrained person will pick the wrong plant. Mushrooms are especially difficult to identify from descriptions in books. Usually, the recipes caution an inexperienced person to avoid the use of certain potentially dangerous plant materials. A cautious person may not want to attempt making certain medicines and may seek a folk healer with an established reputation.

Today, those who practice folk medicine use many natural plant products to treat their patients. Medicinal plants are part of a health-care tradition that has been passed from generation to generation. Because of this, the various powders, liquids, and ointments made from these plants are frequently called traditional medicines.

Traditional medicine began with the first human communities. Over time, humans learned about the food and medical value of plants by using a trial-and-error method. Wrapping a bleeding wound with large leaves might have been the first therapeutic use of plant material. When this primitive bandage proved to be successful, it became a standard method to treat wounds. An observant clan member might have noted that certain plant leaves speeded healing. Eventually, that plant might have been given a distinctive name and designated to be used as a dressing for injuries. The discovery of a beneficial plant is a positive outcome of a trial-and-error experience.

However, the use of the wrong plant might be fatal. When such an error resulted in a death, clan members would attempt to identify the cause. Eventually, perhaps after a series of similar deaths, the poisonous plant would be identified and the knowledge of its effects would be passed on to others.

Humans often assigned special powers to deadly plants. They reasoned that if the full power of the plant could sicken and kill a human, perhaps a small amount of the plant's power could destroy an illness. Further trial and error sometimes led to a successful use of such potent materials. Today, those scientists who search for medicinal plants often use a similar strategy. They also believe that a plant with a strong biological effect —even a poison—might have medicinal properties.

In their search for foods and remedies, humans found the consumption of some plants to be disagreeable but not deadly. Others were found to cause mild stomach upsets or vomiting, while still others caused severe cramps and violent intestinal problems. People learned to avoid these unpleasant plants or used them to purge—or rid—the body of more deadly materials.

In addition to plants that acted as purges, people in preindustrial societies found that certain plant materials relieved pain, provided unusual energy, or caused visions of strange colors and objects. These people reasoned that if plants could bring on such beneficial or extraordinary effects, the plants might be gifts from the gods. The use of such plants in religious ceremonies developed. They continue today among some groups. Participants in these religious ceremonies—such as those followed by some clans in the Amazon regions—ingest plant material to induce visions. The worshipers believe that such revelations expand their religious experiences.

Traditional healers have long been guided by a set of ideas as they select their medicinal plants. Some of these ideas include the principle of "signatures." Healers believe that if some part of a plant resembles an organ of the human body, the medicinal value of that plant will be directed toward that specific organ. In other words, if the flower of a plant resembles a heart, medicine made from the plant is thought to bring relief for heart disease.

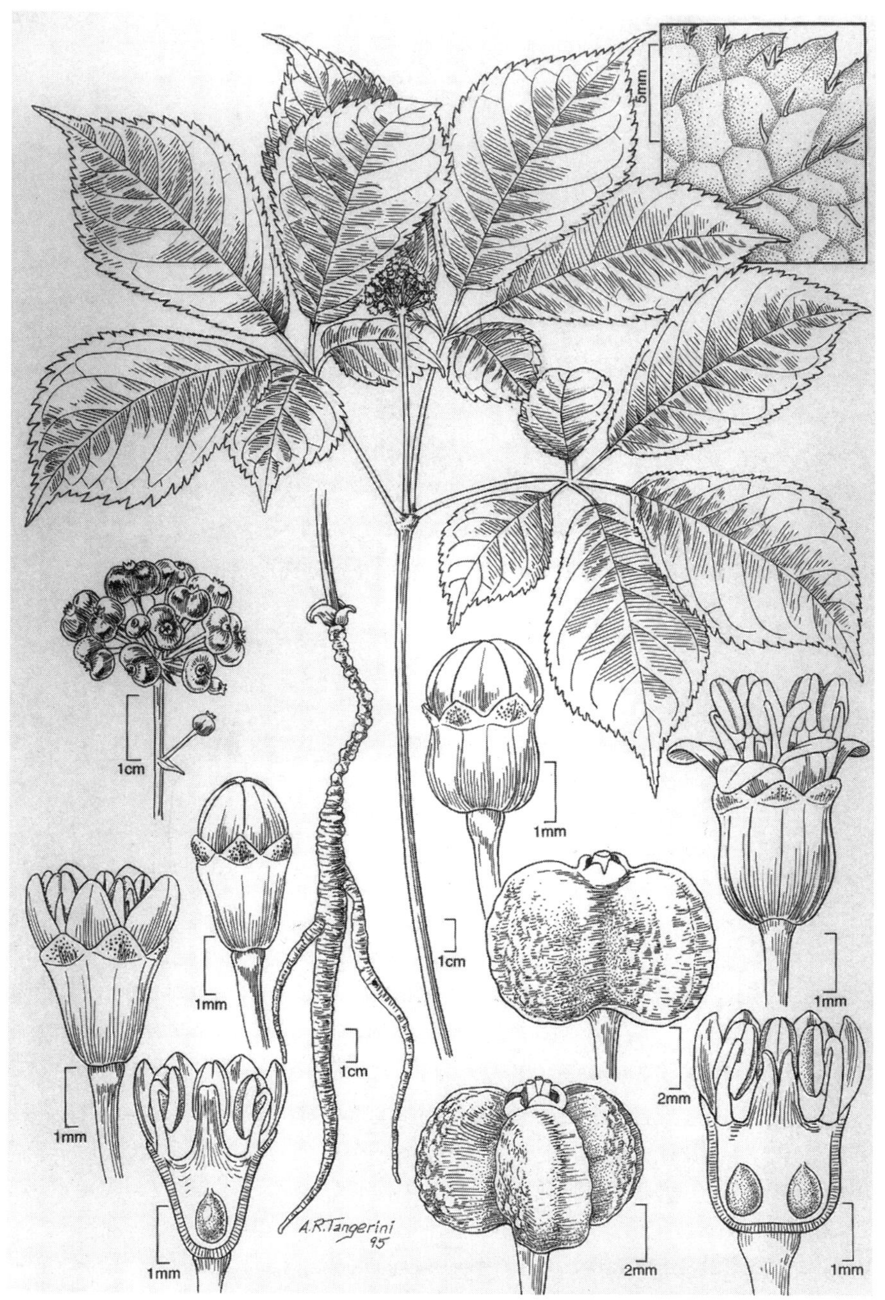


*Extracts from the ginseng root have been used for centuries as a general tonic, first in China and now all over the world.* (Courtesy of Alice Tangerini and the Smithsonian Institution)

This theory is common throughout the world. For many centuries, the Chinese have regarded the root of the ginseng plant as an important medicine. This stems from the idea that the root resembles a human body in miniature. Consequently, powdered ginseng root is thought to increase the health and vigor of the entire body. Today, the use of powdered ginseng has become popular in Western cultures.

Another kind of signature plant is thought to resemble the cause of an illness and, therefore, is used to cure that illness. One example of this is the use of a species of the *Rauwolfia* plant, whose root looks like a snake, to cure snakebite.

Some folk healers believe that the area in which a disease is prevalent supports the plant that cures the disease; this is known as colocation. Consequently, people were not surprised when the cinchona tree was found in tropical regions where malaria is common. The bark of this tree produces the medicine quinine, which relieves the fevers of malaria, a tropical disease transmitted by the *Anopheles* mosquito.

In this case, the idea about location seemed to be verified. However, the effectiveness of a medicine has nothing to do with its appearance or the place where it grows. Indeed, for many years, the cinchona tree would not thrive in India, the region where malaria is most prevalent.

Although signatures and the idea of colocation proved to be invalid, the trial-and-error method uncovered many natural medicines. Over the centuries, positive results occurred because healers were willing to test almost every plant, note the successes and failures, and pass on their findings to others. This persistence —and some good luck—has given folk medicine practitioners many treatments that are still in use.

Today, scientists often consult with tribal shamans when seeking new medicinal plants in regions such as tropical rain forests. They also study the lifestyles of preindustrial people to gain information about local plants. The discovery of curare in the rain forests of South America is a classic example of such a study. In this area, native people hunted with blowguns and curare-covered darts. Curare, a poisonous drug made from a vine, paralyzes its prey. Explorers saw the effects of curare as a possible treatment for severe agitation—such as an epileptic seizure. Although that usage was found to be impractical, curare later proved to be a valuable muscle relaxant when employed in anesthesia. In general, if a plant exhibits any intense biological effects, scientists attempt to isolate and study the active ingredients as a possible treatment for disease.

In the practice of folk medicine, ideas such as plant signatures are shared across cultural boundaries and over time and space. However, the great diversity found in traditional medicine is caused by differences in local diseases, native plants, and cultural practices.

## The Influences of Native Americans

The idea of rigorous tests for health-care procedures was not introduced until the early 1900s. In the 1500s, during the first exchanges in North America between native peoples and Europeans, medications and treatments used on both sides of the Atlantic lacked any type of control. European practitioners often used in their medicines what many would consider truly unpleasant ingredients—such as dried excrement and pulverized sexual organs of animals. Similar materials were employed by the native peoples of North and South America.

The story of an early cultural encounter illustrates the nature of the exchange between two prescientific cultures. In late 1535, the French explorer Jacques Cartier began his expedition up the St. Lawrence River of what is now Canada. Because of various delays, he did not reach the vicinity of present-day Montreal until winter had arrived. After a few weeks, his ships were immobilized by ice. The crew was soon low on rations and began to show the symptoms of scurvy, a life-threatening vitamin deficiency. At that time, no one understood the cause and treatment of this disease. One hundred men—out of the crew of 110—were unfit for duty by February 1536.

The diary of one of Cartier's crew members relates that a local Iroquois leader became very ill from scurvy. The Frenchmen were surprised when the man appeared to be fully cured a few days later. Cartier observed that the Iroquois women brewed a mixture of juniper needles and bark and gave the tea to those who were ill. The dregs of the brew were used to bathe the legs of the afflicted men. When the Frenchmen were offered this treatment, they disliked the strong smell of the tea and at first refused to try it. Finally, some consented to drink the concoction and were soon relieved of their symptoms. Although the sailors did not understand the reason, the scurvy was cured by the vitamin C found in the needles of the juniper tree. The Europeans had unknowingly obtained the needed vitamin by drinking the brew. Bathing the legs did nothing for the sufferers—except make them smell like a pine tree.

This story is often told as one example of an effective folk medicine. Although the Iroquois did use the juniper mash in a medically effective manner, this remedy appears to be another outcome of the trial-and-error method rather than a treatment derived from a true understanding of the illness.

Both the Iroquois and the French thought scurvy was a disease transmitted by physical contact. The French medical procedure was simply to avoid being near anyone who showed the symptoms. The Iroquois hoped to relieve the ill effects of the contagious disease by serving the juniper tea. Neither the French nor the Iroquois had a true idea of the cause of the disease or the reason why the treatment worked.

Native American medical practices, such as the medicinal use of juniper needles, varied from tribe to tribe and region to region. Many of these practices were closely allied to religious convictions that included a belief in supernatural powers. Indeed, the native priest, or shaman, often recited incantations or prayers over the body of the patient. This alliance between medicine, superstition, and religion was similar to European concepts in the 16th through the 19th centuries.

During the early 1700s, some English colonists adopted the use of local medicinal plants when they were unable to obtain their usual medicines. For example, they followed the Indians' lead and used tobacco as a cure for various ailments. Ironically, the conditions so treated included lung congestion. In contrast, other settlers did not trust the Native American remedies and made their own preparations from local medicinal plants. The Puritan settlers in New England went even further. They refused to use any local remedies and imported all their curative ingredients from London.

While some spread of knowledge and practice did occur, neither Europeans nor Native Americans shared all of their health-care information. The prejudices of the settlers and the reluctance of Native Americans to reveal the details of their tribal practices prevented much cultural exchange.

The Native American contribution to modern medicine is difficult to evaluate. However, it is helpful to note the actual number of Indian remedies accepted by physicians in the United States. In the 1800s, United States Pharmacopoeia, an official list of drugs made from medicinal plants, included 200 Native American remedies. When the new laws on safety and effectiveness were put into practice in the 1920s, many traditional remedies—including many of Indian origin—were removed from the list. Today, only 13 North American Indian remedies remain in official use. Even garlic—a time-honored medicine —was dropped in 1936. Interestingly, processed garlic is now a popular food supplement that some believe has health-giving properties.

Of the Indian remedies accepted in the 1800s, a few powerful laxatives and some rubs and ointments are still in use. Perhaps the most valuable are wintergreen and witch hazel, important external medicines for soothing aching muscles. Other native materials still in use have undergone radical changes in function. Cornstarch, once used as a remedy for poisoning, is now a component in bath powders. Wild cherry extract, a flavoring agent to hide the unpleasant taste of strong medicine, was previously a popular cough remedy.

Plant remedies from South America have also been used by traditional folk healers on both American continents. Products

*American Indians used tobacco primarily in ceremonies. Europeans adopted it as a health aid and later as a recreational drug.* (Courtesy of the Photographic Archive Services of the National Library of Medicine)

*The ipecac plant's root produces a juice that causes severe vomiting. Combined with sugar water, it makes a syrup that can be given to young children to induce vomiting. This is useful when a child has swallowed a substance accidentally.* (Courtesy of the New York Botanical Garden)

from some of these plants—kola nut extract, cayenne pepper, and ground arrowroot—are now used as flavorings in food or drink. Powdered papain, once an important drug, is now used as a meat tenderizer. Today, cochineal, made by drying and grinding certain insects, is a red coloring agent. However, four significant South American drugs continue to be used as internal medicines. These are quinine, cocaine, curare, and ipecac.

These four medicinal substances made from South American plants have been important to modern health care. Quinine, the first effective drug employed in the treatment of malaria, has saved millions of lives in many regions of the world. (Although it has now been replaced by other, more powerful malaria drugs, for years quinine was the only treatment.) In some places, cocaine is utilized in eye examinations to dilate the pupil of the eye. Curare is an ingredient in anesthetics. The powdered root of the ipecac plant is used in a syrup to induce vomiting. The syrup can be safely administered to children who have swallowed poison.

While a few medicinal plants from North and South America are still used to cure or alleviate major health problems, many of the natural remedies used in the past have been found to have little real merit. Nevertheless, during the 20th century, members of the scientific community as well as practitioners of folk medicine continue to search for medicinal plants that will help the sick.

## Modern Folk Practitioners

In recent years, scientists have sought out and interviewed folk practitioners in the United States and shamans in remote places such as the Amazon jungle as part of their search for new medicinal plants. The scientists hope to acquire information about unknown medicinal plants and health-care practices before this knowledge is lost in the press of modern life.

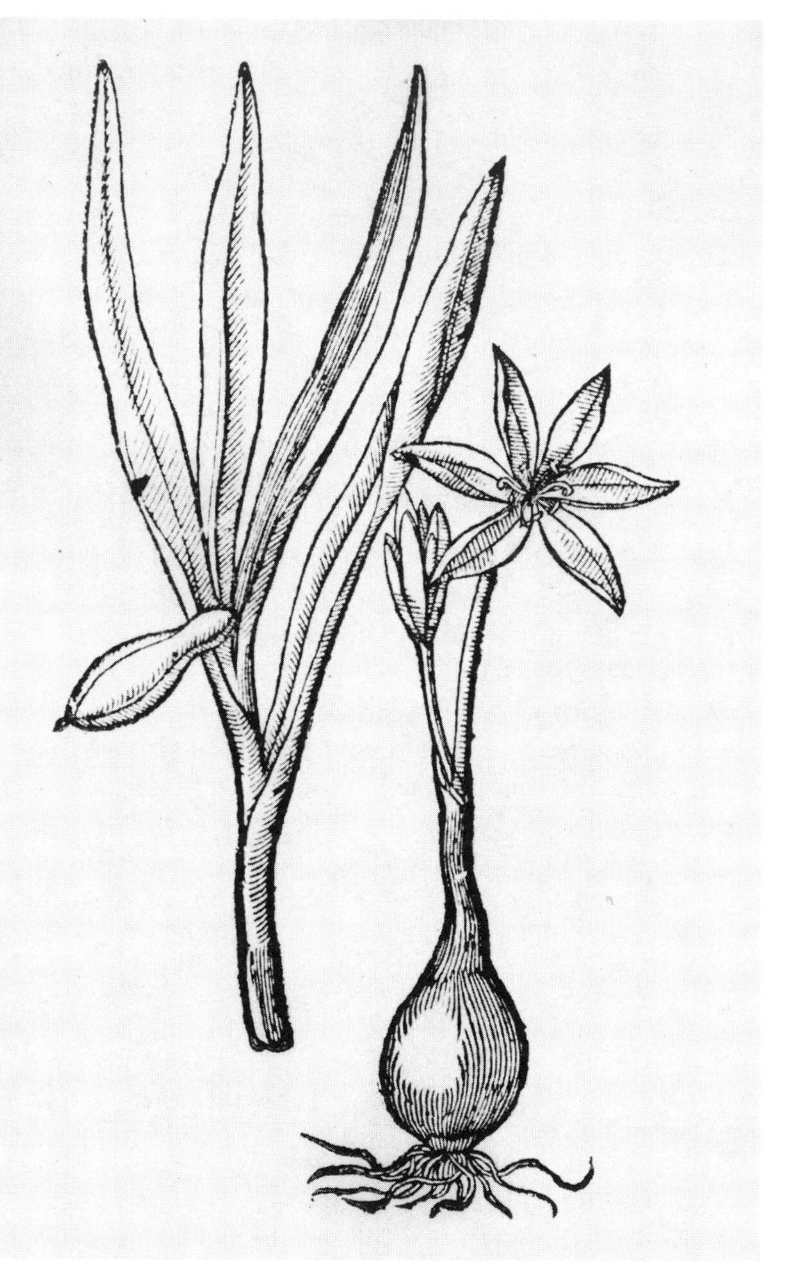

*Although the meadow saffron plant is poisonous to humans and animals, the juice from its seeds and roots has been used to relieve pain and as a specific treatment for gout, a disorder of the urinary system.* (Courtesy of the Photographic Archive Services of the National Library of Medicine)

## Integration of Old and New Practices

In Western Europe, traditional medicine is more integrated into mainstream health care than it is in the United States. In France and the United Kingdom, the use of herbal remedies is an established practice. In these countries, the effectiveness of such materials is not tested by the government. Physicians prescribe and pharmacists dispense traditional remedies that have been in use for many years. There is still a strong traditional element in these practices.

Before herbal compounds can be dispensed as medicines in Germany, they are tested for safety and effectiveness. However, many of these products cannot be sold as medicines in the United States. A prime example is the genus *Echinacea*. The juice of this plant is said to contain chemicals that may activate the body's

*Related to the black-eyed Susan coneflowers, echinacea is used to strengthen the body's defenses against infections of all kinds.* (Courtesy of James Manhart, Texas A&M University)

own defenses against many different infections. *Echinacea* is sold as a dietary supplement—not a medicine—in the United States.

Other products popular in Germany include ginkgo extract, chamomile tincture, and the powdered leaves of the feverfew plant. Many similar materials have been approved for health care by a German government commission.

In Germany, the commercial value of such health-care products has become increasingly important. Indeed, the cultivation of the herbs, the extraction of the active ingredients, the packaging of the products, and the marketing of the remedies are directed by industrial firms that specialize in this technology.

These products are designated as dietary supplements in the United States. Commercial trade in dietary supplements is now the principal source of herbal medicines. The U.S. Food and Drug Administration (FDA) oversees some aspects of this commercial activity such as product labeling. The FDA regulates dietary supplements by requiring accurate labeling and restricting unproved health claims.

An agency of the United Nations, the Food and Agricultural Organization (FAO) is active in international negotiations that are concerned with dietary supplements. The organization works toward international rules and regulations for these products that are in agreement with the rules and regulations of each member country. In addition, the FAO promotes identical labeling practices on all products sold as dietary supplements in international trade.

4

# How Medicines Work

Medicines are chemical compounds. Many people are disturbed when they hear the word *chemical* because they think of poisons and pollutants. However not only are all medicines chemicals, all foods, natural as well as synthetic, are made of chemicals. In fact, all human bodies are made entirely of chemicals. Without chemicals there would be no medicines, no food—and indeed, no life.

The basic unit of chemistry is the atom, a tiny particle of matter that cannot be further broken down and still remain that element. At present, chemists have identified 94 naturally occurring elements and 14 synthetically formed types. These 108 different atoms are the 108 elements that make up everything on the earth. Each of the elements has been assigned a special name and an abbreviation of that name.

Two or more atoms joined together form a molecule. Molecules may be composed of atoms of one single element or atoms of two or more different elements. For example, oxygen (O) is an element. The air humans breathe is composed of the molecular form of oxygen, which is two atoms of oxygen joined together ($O_2$). Water molecules are made of two atoms of the element hydrogen ($H_2$) and one atom of element oxygen (O). The chemical composition (formula) of water is written as $H_2O$.

The small number, or subscript, refers to the number of atoms of each element in the molecule.

Except for water and a few other materials used in very small amounts, all molecules made and used by living organisms contain the element carbon (C). Carbon, like oxygen, can bond with itself ($C_2$). When it does so, it forms a crystal-like structure that does not support life. Both diamonds and graphite, the material used as pencil lead, are carbon in crystal form. Life-giving, carbon-based molecules always include other elements such as hydrogen and oxygen. Most medicines contain carbon-based molecules. These carbon-based molecules often include other elements such as nitrogen (N), sulfur (S), iron (Fe), and magnesium (Mg).

## The Cell

The tissues and organs of plants and animals are composed of thousands of tiny cells. The cell is the basic unit of biology. Each cell has a wall or membrane that keeps it intact and separate from other cells. The cell is the place where diseases strike. For example, infectious diseases may occur when microbes such as bacteria or viruses invade the cell. The action of the microbes can kill the cells, and the death of the cells may result in the symptoms of the disease.

In some diseases such as diabetes, the cells of particular organs fail to function properly. Specifically, in diabetes, cells in the organ called the pancreas fail to produce an enzyme called insulin that keeps the proper level of sugar in the blood. Some diabetic patients must add insulin to their bloodstream by injection.

When body cells reproduce so rapidly that their growth prevents other cells from functioning, the condition is known as cancer. The onset of cancer means that something has gone wrong with the chemical processes within the cancerous cell.

Living cells are remarkably active chemical factories. In the space of a few minutes, a living cell can disassemble and reassemble thousands of molecules. The chemical action of a cell causes the atoms within the cell to group together to form new parts for that cell or for other cells. The cessation or disruption of this work is a basic cause of disease.

The construction of sugar molecules is the main activity of plant cells. Sugar molecules are made from carbon dioxide, which is composed of one carbon atom and two oxygen atoms, ($CO_2$) and water ($H_2O$). Plant cells use the sun's energy to break apart the molecules of carbon dioxide and water in order to build sugar molecules. The plant captures some of the sun's energy and stores it in each molecule of sugar.

After the sugar molecules are manufactured, some are joined together in chains to form starch. In turn, some of the starch chains link together to form cellulose. The energy in the sugar molecules is available to the plant for immediate use. However, the energy in the starch chains is stored for future use. For example, the starch stored in a seed will supply the energy for its underground growth. The longer cellulose chains are employed as the raw material for the assembly and repair of cell walls.

Plant cells manufacture other materials in addition to sugar, starch, and cellulose. The simplest compounds are vegetable oils. Some oils are long, straight chains of carbon and hydrogen atoms. In other vegetable oils, these straight chains have added side chains of various lengths.

Some substances are formed of carbon atoms joined in a circle or ring. The simplest of these ring-shaped molecules is called benzene. In 1825, Michael Faraday, a British scientist, isolated this compound when he was distilling a residue of whale oil. The benzene did not react with other chemicals in the expected manner. Faraday was curious about the unusual material and studied it further. However, he was unable to determine the exact structure of the molecules.

In 1865, the German chemist Friedrich Kekule attempted to identify the true structure of benzene. While deliberating on this

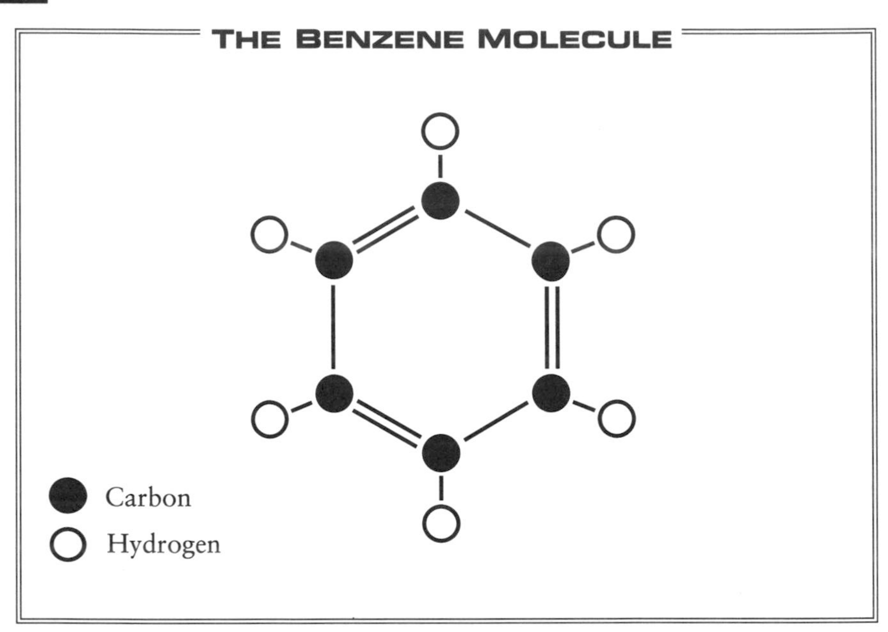


*The benzene ring molecule is formed by six carbon atoms. One hydrogen atom is attached to each carbon atom. This molecule is a building block for many important medicines.*

problem, he supposedly had a prophetic dream. It is said that he dreamed of six snakes, each biting the tail of the one ahead and forming a perfect circle. Whether this story is true or not, Kekule later proved that benzene is a flat ring of six carbon atoms. Today, scientists know that benzene rings are key components of many important plant medicines.

Although carbon, oxygen, and hydrogen are essential to all living things, other elements are also important to life. Some of the most interesting molecules are composed of nitrogen attached to carbon rings or other carbon structures. These molecules are called proteins. Plant cells build protein to serve many purposes, such as binding together the cellulose chains to form the cell walls.

A special class of proteins is called enzymes. In addition to nitrogen, enzymes may contain sulfur, magnesium, copper, or iron. Enzymes are catalysts, substances that activate or regulate chemical reactions but are, themselves, unchanged by the reactions.

Chlorophyll, the green plant enzyme, contains magnesium and utilizes the sun's rays to produce sugar. All plant leaves are filled with chlorophyll, and most have acquired the green color of the enzyme. The amazing manner in which the magnesium in leaves helps to capture the sun's energy remains somewhat mysterious.

The element iron is in the enzyme hemoglobin, a component of red blood cells. Iron is a vital ingredient because it absorbs oxygen from the air we breathe. The hemoglobin then carries the life-sustaining oxygen to all parts of the body.

Enzymes activate many important processes in cells—such as the complicated production of proteins. To do this vital work, enzymes must link several specific molecules. Without the enzyme, the molecules would float past one another in the liquid inside the cell. To initiate protein production, the enzyme fits into one of the specific molecules like a key into a lock. The protruding end of the "key" acts as a hook to catch a different specific molecule; after this linkage, the reaction between the two molecules can take place.

## Enzymes and Medicines

Diseases can be caused by a shortage of a particular enzyme. In some cases, a medicine can substitute for the missing enzyme and help cure the disease. For example, in the disease called epilepsy, there is a shortage of enzyme that control the connections between nerve cells in the brain and nerve cells that control the muscles. For many years, scientists have searched for the correct mixture of enzymes that will prevent the symptoms of epilepsy without reducing the normal activity of the brain. Their goal is to find a replacement for the missing enzymes.

Some diseases are caused by bacterial infection. These infectious bacteria produce foreign enzymes after invading the body. Medicines can link with the foreign molecules, interfere with the function of the invading microbes, and limit the effects of the disease.

## The Chemical Source of Plant Medicines

Some plant cells produce strange enzymelike molecules that do not contribute to the productive work of plant cells. Since scientists tend to believe that everything in nature has a purpose, they are curious to discover the use of these unusual enzymes. They now believe that the strange molecules might be part of a plant's defense system. Some of the molecules have a bitter taste and others are poisonous. Perhaps these mysterious chemicals help the plant ward off its enemies. Scientists hope that some of these bitter or poisonous enzymes can be used as medicines.

Today, medical researchers are able to adjust the dosage of a medicine containing poisonous enzymes so that only selected cells are killed. Scientists can also redesign the molecules of these enzymes to kill cancer cells but allow healthy cells to survive.

## Specific Cases of Poisons Transformed into Medicines

Scientists are investigating particular chemicals that possess poisonous properties. For example, taxine is a material from the leaves of the western yew tree. This substance is highly toxic when absorbed through the skin. This poisonous property made scientists interested in the plant. Unfortunately, they found that taxine had no value as a medicine. However, the bark of the western yew manufactures a different chemical called taxol. This chemical, discovered in 1971, may be produced by the tree to defend it from bark-burrowing insects. In 1977, scientists discovered that taxol is a highly effective medicine against various cancers including ovarian cancer. At the proper dosage levels, the medicine kills cancer cells but leaves healthy cells intact. Unfortunately, the western yew is relatively rare, and the tree population is threatened because the removal of the bark can kill the tree. Work on a synthetic form of taxol may reduce this threat.

Native healers in Jamaica have long used an extract of the periwinkle plant to treat diabetes. In the mid-1950s, Canadian scientists investigated the active ingredients of the plant. At about the same time, U.S. researchers learned that folk healers in the Philippines used a similar substance to treat diabetes. Both the U.S. and the Canadian research teams soon found that periwinkle chemicals did nothing to alleviate diabetes. However, the U.S. team determined that vincristine and vinblastine, two of the enzymelike molecules produced by the plant, kill the cancer cells of leukemia.

*The green needles and red flowers of the western yew tree are highly poisonous. So is the inner bark, but it is the source of taxol, a medicine that is used to treat cancers.* (Courtesy of James Manhart, Texas A&M University)

*Extracts from the rosy periwinkle plant provide two chemicals that are important cancer-fighting drugs.* (Courtesy of James Manhart, Texas A&M University)

## A Range in Effectiveness

There is a constructive role for medicinal plants in medicine. However, the misuse of these plants could undermine the honorable status of legitimate herbal medicine. For example, the claim that a plant can cure a wide range of ills or relieve inherited diseases such as epilepsy or diabetes is highly suspect.

Potions intended to be curative can be both ineffective and unsafe. Today, scientists know that sassafras oil and a related product, safrol, can cause cancer in laboratory animals. However, some people continue to recommend these materials as home remedies. They are prescribed for colds, fevers, and stomach upsets and are served as a tea brewed from pieces of the root.

Some medicinal claims defy common sense. For example, a tea made from the bracken fern is recommended as a treatment for lung ailments. This fern, which grows in stagnant water, is known to poison grazing cattle. While such a plant poison might

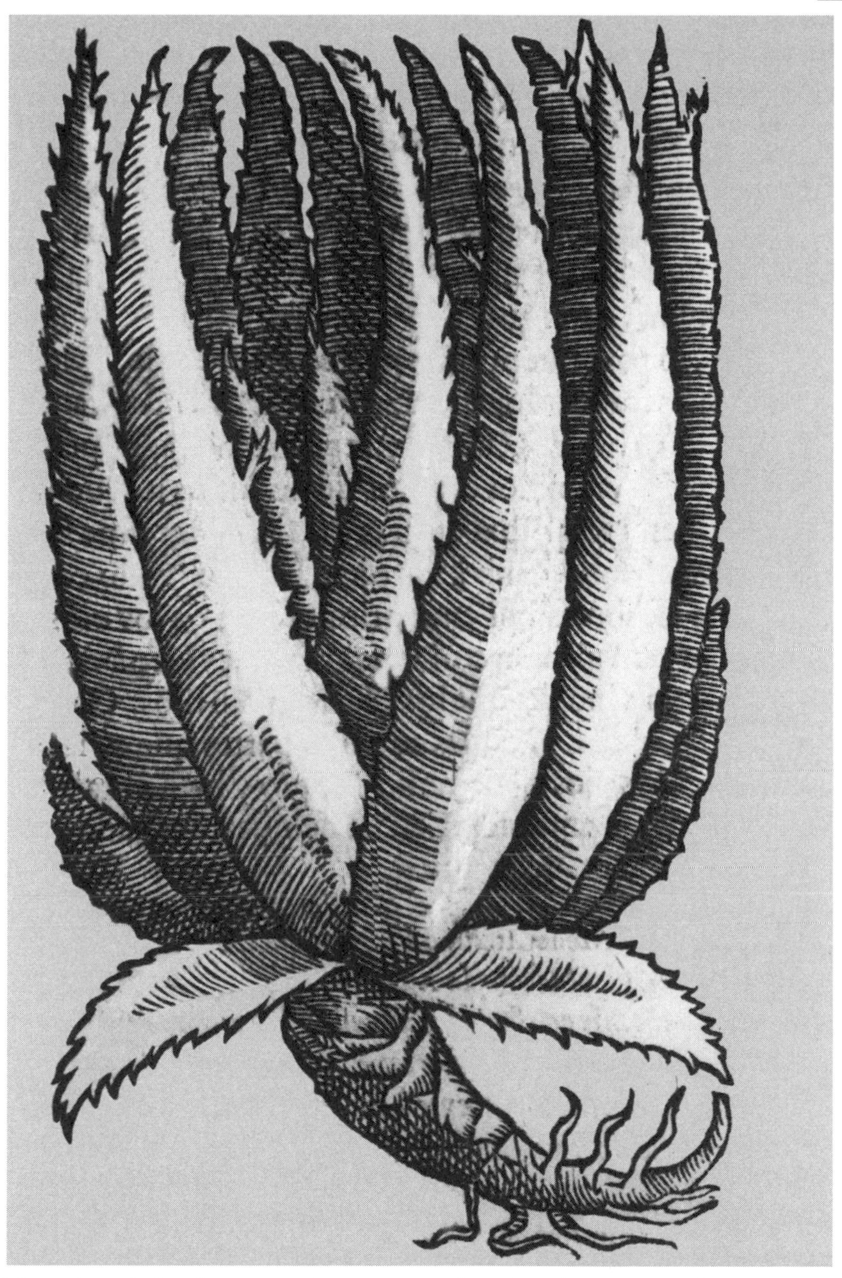

*The juice of aloe leaves makes a strong laxative that can weaken a patient. When the juice is used on the skin, however, it can relieve itching and redness.* (Courtesy of the Photographic Archive Services of the National Library of Medicine)

be refined into a true medicine, the raw form is unlikely to be beneficial and could kill the unwary.

Some traditional herbs are modestly effective. The active ingredient in thistle sap can promote the healing of wounds. Some herbalists also recommend thistle sap to be brewed as a tea. However, the tea is ineffective because the active ingredient is not water soluble.

An extract of *Echinacea*, a cousin to the black-eyed Susan, is useful in enhancing immunities against viral diseases. However, despite claims to the contrary, *Echinacea* is useless in treating tetanus infections or poisonous insect and animal bites.

Raw garlic has some effect against viral infections but not if the smelly but active ingredient, allicin, is removed.

Although most medications are useful for only one illness, a few can combat more than one. Cortisone, a hormone now made from yams or soybeans, is one of those rare medicines. It is used successfully against arthritis, allergies, skin rashes, and several other medical problems.

Aloe is a plant credited with curing many ailments including both stomach and blood problems. However, if ingested by mouth, aloe acts as a strong and unsafe laxative. The plant is also recommended for many skin problems and is often mixed with cortisone—the proven skin remedy. Used externally, aloe seems to be beneficial.

Today, most folk remedies are not marketed as medicines in the United States. This marketing strategy avoids the requirements imposed by the U.S. Food and Drug Administration on products specifically designated as medicines. Instead, the folk remedies are sold as food products or dietary supplements. Producers of herbal teas, for example, can claim that the teas will bring a wide range of health benefits. However, they are not required to indicate which ingredients have specific medical effects. While they cannot claim that the materials will cure specific diseases, they can use words such as *restorative* that have very general meanings.

# 5

# The Regulation of Medicines

By the late 1700s and early 1800s, many shady business people in the United States were manufacturing health-care products containing medicinal plant materials. Few had credentials in science or medicine. Many successfully promoted their products by presenting dramatic tent shows, staging colorful parades, and placing large advertisements in newspapers.

Their remedies became known as "patent medicines." A patent is legal document that gives the inventor the sole right to produce and sell an invention for a set period of time. In the case of these remedies, the name of the product could be protected by the trademark law. However, the remedy and its formula could not acquire a legally binding patent unless the inventor revealed the ingredients. In those days, few producers wished to disclose their secret mixtures. So, ironically, most early "patent medicines" were not really patented.

The first American patent medicine was marketed in 1796 by a man named Samuel Lee from Connecticut. Like other products of the time, "Lee's Bilious Pills" were advertised as a cure for many different diseases. His pills were supposed to cure yellow fever, jaundice, dysentery, dropsy, worms, and "female complaints" as well as biliousness (heartburn and gas). The name of the product, the only thing protected by the "patent" (really a trademark), was widely advertised. However, the trademark was quickly violated by three other people named Lee. Each marketed his own "Bilious Pills." The three other Lees profited from the advertising financed by original marketer, Samuel Lee.

Members of the medical profession attempted to control or suppress the distribution of patent medicines through their own advertising, public crusades, and lawsuits. The first governmental attempts to curb the marketing of these products began in the late 19th century with laws to correct false or incomplete labeling. The manufacturers were ordered to adhere to the official rules of the professional association of pharmacists. Patent medicine labels had to list both the ingredients and the amount of each ingredient used in the remedy.

These efforts to control patent medicines were countered by some newspaper owners who earned large sums of money for advertising the products. Newspaper proprietors and patent medicine manufacturers were allies because of the financial gains for both.

In those early days, manufacturers of many health-care products made exaggerated and unproven medical claims about their remedies. Thousands bought their products. Even today, people who cannot be helped by conventional medicine sometimes turn to products making unfounded claims and buy questionable remedies. As long as there are difficult or incurable diseases, there will be a market for controversial health-care products.

## Other Dangerous Products

While some patent medicines were worthless but harmless, others were dangerous and habit forming. Some of the most important habit-forming drugs come from plants; an example is the opium poppy, a plant native to the Near East. For thousands of years, opium and opium derivatives were considered beneficial for a wide range of medical problems. Abuse of opium derivatives began to draw public attention in the mid-1860s during the latter days of the American Civil War. Morphine, derived from opium, was used for pain relief during and after surgical procedures near the battlefields. Dosages were not controlled, and some patients became heavily dependent on the drug. In many cases, these patients developed a lifelong addiction.

In those times, the financial cost of drug dependency was low. Opiates were readily available through open, legitimate channels and at nominal costs. Sales of these opiates were not restricted in the United States until the Federal Narcotics Act of 1914—sometimes called the Harrison Act. Scholars believe that the peak rate of narcotic dependency in the United States—about 2 to 3 percent of the population—was reached around the turn of the century. Rates of dependency were already declining before the Harrison Act was passed in 1914, on the eve of World War I.

The public became increasingly concerned about the dangers of such opiates during the first decade of the 20th century. They were especially angered by the incomplete labeling found on some patent medicines and elixirs (tonics that claim to improve general bodily health). Many labels did not state that opiates were included in the product. People feared that unscrupulous manufacturers were trying to cause drug dependency to ensure a steady market for their product. Public opinion about such matters finally led to federal legislation.

Harvey Washington Wiley, head of the Bureau of Chemistry in the U.S. Department of Agriculture under President Theodore Roosevelt, worked to reform food and drug practices. Wiley helped Samuel Hopkins Adams, an important critic of U.S. public health laws, to compose a series of articles for *Collier's Weekly*. The series later resulted in the book *The Great American Fraud*.

Wiley led the crusade against the contamination of processed foods. This problem included such common acts as the dilution of milk with unclean tap water. Members of medical and pharmaceutical societies and women's organizations soon joined Wiley's battle. This alliance was the beginning of modern consumer advocate groups.

The patent medicine companies were against Wiley and his allies. Promoters of states' rights, citizens who believed that local and not federal government should hold more power, were also opposed to these regulations. Finally, in 1906, Upton Sinclair published his book *The Jungle*. Sinclair, an American writer and reformer, had written about the unsanitary conditions of the Chicago stockyards. His account of the greedy people who produced dangerous food products led to a huge public outcry. The popularity of the book forced President Theodore Roosevelt to propose some regulations for the food and drug industries. After much wrangling and many attempts in Congress to kill or weaken the bill, the Pure Food and Drug Act was signed into law on June 30, 1906.

The law resolved some troubling aspects of the problem. The new law prohibited the shipment of contaminated materials across state lines. Marketers of patent medicines were forced to obey a minor "truth in labeling" provision. However, the law did not require manufacturers or sellers to test the effectiveness of their product or warn users of potential side effects. Most producers quickly complied with the labeling requirements. By following the new rules, they hoped to regain the confidence of their consumers and ensure their future prosperity. Indeed, their customers returned.

The government and the crusaders, however, had just begun to fight. In 1912, the law was revised to provide penalties for false therapeutic claims. The enforcement role was assigned to the U.S. Department of Agriculture (USDA). By 1914, control of narcotic materials was brought under federal jurisdiction by passage of the Harrison Act. Five years later, the enforcement of these laws was given to the Food and Drug Administration (FDA), a new organization within the USDA.

In 1937, a major scandal arose when a drug company mistakenly mixed a large batch of antibiotic medicine with a syrup containing a poisonous compound. Several hundred people died from this error. The next year, the U.S. Congress added safety testing to the FDA's responsibilities. No new drug could be put on the market until the producer completed the tests listed on the new drug application form.

Laws passed in 1952 made a sharper distinction between proprietary drugs that could be sold over the counter and prescription drugs that were too dangerous for self-medication.

A tragic episode during the early 1960s initiated the next set of reforms and regulations. Thalidomide, a medicine developed and approved in Europe, was intended to ease upset stomachs. In Europe, the nonprescription medicine was given to pregnant women suffering from morning sickness. Beginning in 1961, some of these women gave birth to infants with malformed limbs. However, only a few new mothers in the United States were exposed to this threat. In spite of pressure from an American marketing firm, the drug had not yet been approved for distribution in the United States. When the news broke, the U.S. distribution company's sales representatives had been ready to give samples of Thalidomide to physicians.

In 1962, a new law required drug companies and the FDA to test the safety and effectiveness of all new drugs before they could be released for sale. Also, the 4,000 products that had been approved by the FDA between 1938 and 1962 were restudied by a panel from the National Academy of Sciences. Most of those products came from plants. The panel found that about half the compounds fulfilled their producer's claims, but that 760 were completely useless. The specific medical value of the remaining 1,200 products could not be determined from studies that had been done up to the time of the review.

The effectiveness of a medicinal product is difficult to determine. A compound might appear to work on a few cells in a test tube but prove to be toxic when introduced into the body of a living animal or human. Another compound might be effective in fighting a disease in the body of an animal but fail to work when administered to a human. A drug might appear to produce good effects in some individuals but not in others. Some of the apparently positive effects could be caused by other factors.

Under the rules laid down by the U.S. Food and Drug Administration, the evaluation of a new drug involves at least six stages.

The evaluation process is identical for medicines made from medicinal plants, animal sources, or synthetic materials. The first stage is a laboratory analysis of the basic chemicals in the medicine.

The second test is also done in the laboratory. To test a cancer medicine, for example, the new compound is placed into a test tube of living cancer cells that have been cultivated by a medical technician. The technician evaluates the effectiveness of the new medicine by determining the number of cancer cells that have been killed or prevented from reproducing.

A typical third stage involves the use of small test animals such as rats or mice. For testing cancer cures, these animals are bred to be very susceptible to cancer. When the animal develops a cancerous condition, the new medicine is administered and the fate of the cancerous tissue is observed.

The next stage usually employs a larger animal and determines the safety of the product. Monkeys, with body functions more humanlike than those of rats or mice, are frequently used as test animals.

If all the tests are passed, the next stage will often be the treatment of human patients. Typically, these patients are in the final stages of an illness that cannot be cured by existing medicines. These patients volunteer to be test subjects in the hope of a miraculous cure. The new medicine can be considered effective even if only a small percentage of terminal patients are helped.

Now, the new medicine is ready to be tested on large numbers of human patients. A special safeguard is added to the evaluation at this stage. The testing procedure is called a "double-blind method." For this, the total group of test subjects is divided into three subgroups of nearly equal size. One subgroup is prescribed the new drug. A second subgroup is prescribed a drug already approved to treat the illness. The third subgroup is prescribed a harmless but totally ineffective material called a "placebo." The term *double blind* is used because neither subjects nor testers know which substance is prescribed to each individual since all the prescriptions are coded; both parties are therefore "blind."

In almost every test, a few of the people given the placebo experience some improvement. People who expect to get better often do—even when no curative material has been administered. The new drug will not be accepted by the FDA unless it proves to have a more positive effect than the placebo and a more beneficial effect than the previously approved medicine.

After passing all the tests, the new drug is released as a prescription medicine. However, doctors are notified that the drug is new and that its safety is still uncertain. Physicians are requested to inform the FDA about side effects such as stomach upsets or dizziness. Gradually, if few problems are reported, concern declines and the new drug is regarded as a standard treatment.

At the present time, many new medicines are synthetic copies of natural substances. However, a strong interest remains among health-care specialists in medicines that are derived directly from natural sources—plants and animals living on land or in the water. All new medicinal compounds whether natural or synthetic must pass the tests of the FDA.

6

# South American Expeditions

South America contains some of the world's most extensive rain forests, which have provided humans with several important medicines. Some of this astonishing variety of plant life has been tested for medicinal properties, but many plants remain to be investigated. Today, scientists and environmentalists are fighting to preserve the forests from eradication by commercial interests. Many scientists fear that untested plant species will be destroyed before they can be analyzed. It is hoped new wonder drugs may be found among those unstudied plants.

## Early History of Spanish America

In 1498, Christopher Columbus completed his third voyage to the Americas. He landed near the mouth of the Orinoco River in present-day Venezuela. Columbus's journeys had been financed by the Spanish monarchs, Ferdinand and Isabella, and Spain soon led the world in both exploration and colonization. During the next 10 years, Spaniards explored the east coast of South America. In 1513, the Spanish captain Vasco de Balboa trekked across the narrow Isthmus of Panama. He was the first European to stand on the eastern shore of the Pacific Ocean.

In 1527, another Spanish captain, Ferdinand Magellan, entered the Pacific by sailing around the southern tip of South America. Francisco Pizarro duplicated that route to the Pacific, and in 1535 established the city of Lima, Peru. Other Spaniards settled at Cartagena on the Caribbean coast of Colombia. Their countrymen also occupied Quito, a city previously held by the Incas in a high Andean valley near the equator.

In pre-Columbian times, the Inca people lived along the Pacific coast and in the central regions of the Andes Mountains. The highly developed Inca civilization occupied an area with many gold and silver mines. The Spanish invaders saw evidence of these riches in the Incas' artifacts—their statues, jewelry, and other objects made of gold and silver. In their quest for wealth and power, the Spanish seized the artifacts, exploited the mines, enslaved the Inca, and helped destroy the already declining Inca

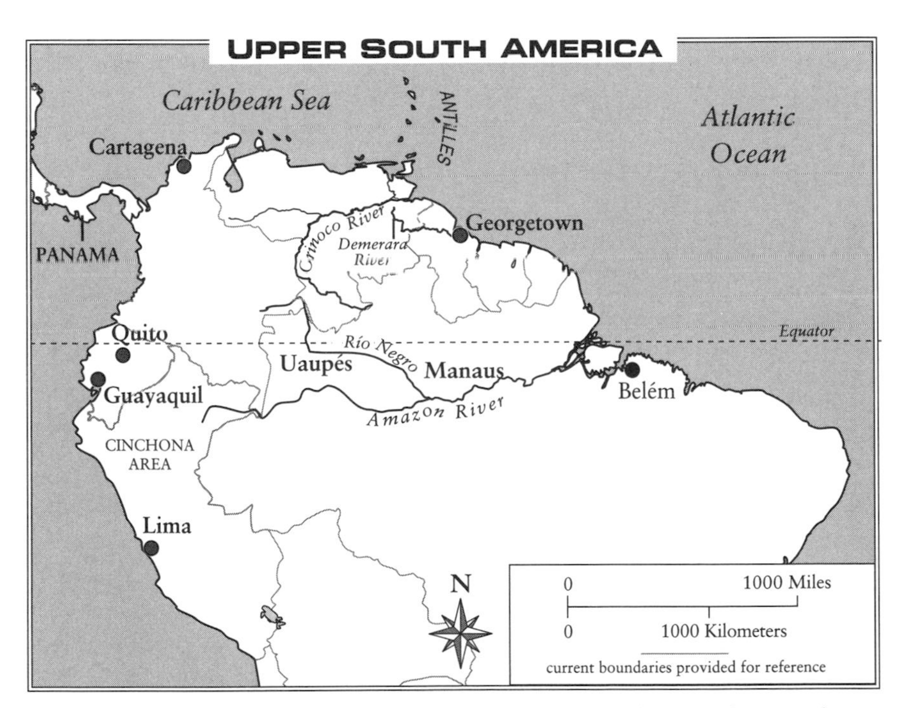


*The map shows the major cities and rivers of the north-central area of South America. The cities served as supply bases and resting places for explorers. The rivers served as their passageways.*

civilization. Attracted by stories of fabulous wealth, Europeans explored the mountainous portions of South America for the next three hundred years.

To ensure a steady flow of precious metals into Spain, the Spanish monarchs retained control over the richest areas of the continent.

## Open-Air Science

Some Spanish and Portuguese explorers had a modest interest in science and recognized that the New World should be studied as well as exploited. However, the financing of research expeditions was given a very low priority by European rulers. These limited resources were partly responsible for the lack of scientific exploration in South America. Other factors, too, slowed the study of the continent's plant and animal life.

Few Europeans were trained as scientists in the time of Columbus. Indeed, there was no clear definition of science or of careers in which scientific knowledge or research played a major role. Some members of the government, the military, and the religious orders were interested in investigating the world around them. This interest, however, was not their main work. Few individuals possessed both the interest and the independent income to allow them to make scientific investigation a full-time occupation. Therefore, in the years between the early 1500s and the 1900s, a relatively small number of individuals were able to investigate the scientific riches of South America.

During that long stretch of time, the people who were able to do such research were known as naturalists. This title meant that the individual had a working knowledge of astronomy, geology, geography, zoology, and botany. Their formal education was often in the field of medicine or the study of philosophy with an emphasis on mathematics. The knowledge and skill required to conduct scientific research was often self taught.

Although some of the individuals who came to South America were concerned with finding medicinal plants, most were interested in discovering new and unusual ornamental plants to beautify the gardens of Spain and Portugal. Most Europeans believed that they had located and perfected all the important sources of medicine. The specific search for medicinal plants in South America is a relatively recent enterprise.

Prior to the late 20th century, expeditions to South America had much broader scientific goals. The early discoveries of medicinal plants in South America were often accidental. Today, the search for medicinal plants combines anthropology and botany. This field is called ethnobotany.

In spite of the informal way in which early explorers sought medicinal plants, Europeans found some useful specimens in South America. Three explorers were responsible for opening the continent to the later scientists who focused on locating medicinal plants.

## Charles-Marie de La Condamine, Immortal of France

Although preceded by a few Spanish naturalists, Charles-Marie de La Condamine of France was the first serious scholar to gain a broad scientific perspective on South America. La Condamine was born in 1701, the son of a high official in the court of the French king Louis XIV. When he was 18, he finished his formal training in mathematics and joined the French army as a junior officer. Ironically, his first active duty was fighting against the Spanish. Between battles, La Condamine conversed with a captured Spanish soldier who had served with the Spanish forces in Peru. The descriptions of the Andes, the South American rivers, and the cities of the Incas fired La Condamine's imagination.

At the age of 29, La Condamine was elected to the French Academy of Sciences. In 1730, shortly after his election, the academy was plagued by a controversy between followers of Isaac Newton, an English mathematician, and Jacques Cassini, an Italian-French scientist. Newton had theorized that the earth was slightly flattened at the North and South Poles because of its rotation and the effects of the moon's gravity. In contrast, Cassini proposed that the earth was slightly elongated toward the Poles. The opposing sides conducted great verbal battles within the walls of the academy. French patriots favored Cassini's ideas because he had French family connections. However, academy members finally realized that the issue was vital to the future of ocean travel and too important to be resolved by debate—they needed evidence.

All the members knew that sailors must be supplied with reliable astronomical measurements in order to navigate their ships across the oceans. The academy began making plans to gain this information. To be absolutely accurate, the measurements had to be made as close as possible to the equator and the North Pole. Within a short time, the northern base was set up in Lapland near the Arctic Circle.

The French soon discovered that finding a good equatorial site was not as easy. Although huge areas of both Africa and South America are crossed by the equator, gaining access to the best possible sites presented formidable problems. The African coast was guarded by pirates who would not allow the scientific expedition to land. In South America, Quito, a city in the Andes Mountains, was an ideal location to make the celestial observations. But Quito was controlled by Spain—a former enemy of France. Nevertheless, negotiations were begun and King Philip of Spain agreed to permit the expedition. The necessary royal commands were signed and the leaders of the French Academy began to organize the expedition to South America. They needed a young, brave, and vigorous scientist to lead the party. They selected La Condamine.

The expedition team consisted of 10 members, including Joseph de Jussieu, an expert in botany. In November 1735, the group landed in the seaport of Cartagena. They were joined by two Spanish naval officers, their escorts for the rest of the trip. Today, Cartagena, a city in northwestern Colombia, would seem an awkward beginning for a trip to Quito, the capital of Ecuador. Then, Cartagena was the official gateway to the Spanish territories. This isolated location was designed to keep people unacceptable to Spain out of Spanish lands.

La Condamine's expedition crossed the Isthmus of Panama and headed south for the coastal town of Guayaquil, several hundred miles from Quito. The group was met in Guayaquil by a local leader, Pedro Maldonado. Maldonado wanted to act as guide to the prestigious group of scientists and told them of his new route to Quito. La Condamine was worried about his men traveling on an untried route and ordered them to use the established way. He alone accompanied Maldonado on the new route.

His route proved rewarding. When La Condamine and Maldonado reached the foothills of the Andes, they were met by natives who showed La Condamine a material made from the sap of a tree. The material, called caoutchouc, was a stretchy clothlike substance. Today, caoutchouc is known as rubber. While not a medicine, rubber has come to play a major role in health care. It has been used for surgical gloves, rubber sheets, tires on movable stretchers, and countless other important objects. Indeed, La Condamine saw an immediate use for the material. He wrapped his scientific instruments in the material to keep them from breaking and then made a container to protect them from the frequent rain. When the Frenchman returned home, he informed his colleagues about this useful plant product. La Condamine is regarded as the modern European discoverer of rubber.

La Condamine and the other members of his expedition were reunited at Quito and soon the group began to survey a 200-mile (320-km) strip of land. This strip would serve as a baseline in their mathematical calculations to determine the size of the earth at the equator. The process took a full year.

*The rough bark of this large rubber tree is sliced open down to the inner layer. The cut is made at an angle so that the raw sap, called latex, will flow down the trunk.* (Courtesy of Marco Bleeker, mbleeker@euronet.nl)

By measuring the angle of a certain star from each end of the baseline, the scientists determined that Newton's theory was correct. The expedition to the Arctic Circle had already confirmed Newton's position. The mission had been accomplished, and the men were ready to return home. La Condamine, however, was not finished with his explorations of South America. He decided to travel down the eastern slopes of the Andes Mountains and then navigate the Amazon River to the Atlantic coast. La Condamine's journey down the Amazon River helped to reveal the vast resources of the rain forests. European residents along the 2,000 miles (3,200 km) of the upper Amazon were mainly Jesuit missionaries sent to convert the natives to Christianity. These missionaries were able to help La Condamine and others survive the rigors of the Amazon region.

Indeed, a local Jesuit priest helped La Condamine enter the river near its highest navigable point in the eastern foothills of the Andes. Other Jesuits along the Amazon told the French explorer about various useful plants that grew in the area. They described quinine and its ability to reduce fevers. They also related the paralyzing and deadly effects of curare and how native fishermen used extracts from the barbasco plant to intoxicate fish. Maldonado and La Condamine are credited with opening up the Amazon River area.

La Condamine recognized the great importance of four South American plants. When he returned to Europe, La Condamine took back samples of rubber and information about harvesting and processing the useful, stretchy material. He carried back seeds and seedlings of the cinchona tree and hoped to cultivate the tree that alleviated malaria. After his return to France, La Condamine experimented with curare and was the first European to note that its effects were not necessarily fatal. He also collected specimens of the barbasco plant, the juice of which was used by natives to intoxicate fish and make them easy to capture. This plant produces saponins, a type of chemical used as a raw material in many medicines. Much later, it was found that rotenone, an extract of the barbasco, is an effective natural insecticide.

## Fusée Aublet

Another French scientist, Fusée Aublet, sailed for South America about 20 years after La Condamine completed his explorations and returned to Europe. Unfortunately, Aublet's life is not as well documented as that of his countryman. Nevertheless, some modern botanists regard Aublet as the founder of ethnobotany. An ethnobotanist obtains information about new plants by consulting with native peoples and records how their local plants are used for food, medicine, clothing, and shelter. Aublet was the first scientist to use this approach.

As a young man, Aublet may have been schooled in medicine as well as botany. He served as a druggist's apprentice in Spain and, briefly, as a medical assistant in the Spanish army. After his army service, Aublet accepted a position as a druggist at the Charity Hospital in Paris, France. He then gained employment with the Company of the Indies, a French colonizing and trading organization. The young scientist was assigned to their base on Mauritius, a large island that lies off the east coast of Africa. Aublet worked on Mauritius as a dispenser of medicinal herbs during the 1750s and early 1760s.

At that time, the French government controlled a territory on the upper east coast of South America that came to be called French Guiana. Fusée Aublet was chosen to inventory the medicinal plants of that area. He arrived at the French territory in 1762 and completed his work in 1764. He collected and identified many new species of plants including 124 with medicinal or nutritional uses.

## Alexander von Humboldt, Liberal Aristocrat

The next extraordinary European explorer of South America arrived on that continent in 1799. This German scientist, Alexander von Humboldt, was born in 1769. He was among those lucky individuals whose family wealth allowed him to be a full-time naturalist. His father died when he was 10 years old and his mother, Marie Elisabeth von Holwede, took responsibility for his education. She had high ambitions for her sons' careers. Marie Elisabeth carefully selected the tutors who were retained to teach Alexander and his older brother, Wilhelm. After leaving home at age 16, his education continued at military schools, and he gained his training as a scientific explorer at German universities. His interest in the arts, languages, and science was due in part to his mother's continuing influence.

After Napoleon's invasion of Egypt, Humboldt, then in his late 20s, was invited to join a French expedition to that country. The proposed journey did not take place because the English fleet defeated Napoleon's forces and blockaded Egypt. After this change of plans, Humboldt and Aimé Bonpland, a botanist who had planned to go to Egypt with him, decided to become partners in research projects. This partnership proved fruitful.

After their hopes to explore Egypt and North Africa were frustrated by the political situation, Humboldt and Bonpland decided to explore South America. Since Spain still controlled access to South America, permission to travel there was required. Humboldt's family connections assured them an audience with the Spanish king Charles IV. Happily, the two scientists received a special passport from the king that secured complete cooperation from the Spanish officials in South America. After careful preparations and interviews with Spaniards who had been to South America, the two left Spain in the spring of 1799. A few months later, they arrived in Venezuela at the mouth of the Orinoco River.

The explorers were eager to confirm the truth of a rumor that La Condamine had carried to Europe. The earlier explorer had been told that there was a natural canal that connected the Orinoco and the Amazon Rivers. Humboldt and Bonpland wanted to locate this connection if it existed. Because rivers were the principal means of transportation through the continent, a connection between the two important waterways could have major economic consequences.

Humboldt and Bonpland traveled first on the Río Apure, a tributary of the Orinoco River. They sailed down the Apure, entered the Orinoco well above its mouth, and then turned upstream on the main river. After many difficulties, they arrived at a plateau in the Venezuelan highlands where the Orinoco divides into slow-moving streams, one of which turns away from the others and flows into the Río Guiana. The Guiana joins the Río Negro, a large tributary of the Amazon. Thus, Humboldt did find a connection between the Orinoco and the Amazon. Sadly, it was not navigable. Instead, it was a shallow, slow-moving creek.

During their river travels, Bonpland had been busy collecting both ornamental and medicinal plant specimens. After confirming the truth of La Condamine's story, the explorers decided to trek to the Río Negro, sail down to the Amazon, and then continue along its course. Bonpland sent his plant collection down the Orinoco with some native messengers and the pair set off overland to the Río Negro.

Much of their route was through Brazil, then under Portuguese rule. The passport endorsed by the king of Spain was no longer valid. In fact, the explorers were promptly arrested as spies when they reached the first military base on the Amazon. It took some time to untangle the confusion. During their return to Venezuela, Bonpland contracted malaria. He was saved from death by large quantities of locally produced quinine.

After some side trips, the pair continued their explorations along the west coast. They sailed up the Río Magdalena to Bogotá, the capital of New Granada (now Colombia). In Bogotá, they met with José Mutis, a remarkable man who combined the talents of physician, priest, and botanist. The two European scientists had studied Mutis's collection of plant specimens in Madrid, Spain, before leaving Europe for South America.

While in the western area of South America, Humboldt began a series of studies that concerned both geography and botany. He focused on climatic variations and the relationship between plants and the temperature and altitude in which they grow. Such studies allowed Humboldt to explain why plants from low-lying areas of cool regions have strong similarities with plants that thrive at high elevations near the equator. This observation was important for those who would later attempt to domesticate medicinal plants.

Bonpland was particularly interested in collecting plant specimens from a certain area around Loja in Chile. According to legend, quinine from trees in this vicinity allegedly cured Countess Chinchón, a member of the Spanish nobility, in 1638. Her family name was later given to the cinchona (chinchon) tree from which quinine is derived.

While in the area, Humboldt observed the harvesting practices used to remove bark from the wild cinchona trees. He became concerned, as had La Condamine, that the practice would kill the trees and the species would become extinct, eliminating the source of natural quinine forever. Their concerns mirror those of late 20th-century scientists who were worried about the survival of the Pacific yew, a source of taxol.

The explorers chose a coastal route for their journey from Colombia to Chile. Humboldt wondered why the long stretch through Chile and Peru was so arid compared to the coast of Colombia. His knowledge of geophysics provided the answer. The sea off the coast of Chile and Peru is remarkably cold, but the coastal lands are warm. When the cold, moisture-laden winds blow from the ocean, the heat from the land warms the breezes and increases their capacity to carry moisture. Consequently, the air holds the moisture until forced to rise by the Andes Mountains, which lay some distance from the coast. Later, in recognition of Humboldt's activities, the coastal current of cold water, 150 miles (240 km) wide and 1,000 miles (1,600 km) long, was named the Humboldt Current. This stream of water is recognized as the controlling factor in the rich fishing grounds off the coasts of Peru and Chile.

After he returned to Europe, Humboldt was credited with the "rediscovery of South America" and his adventures generated great curiosity about the continent. Shortly after Humboldt returned home, the peoples of South America overthrew their Spanish and Portuguese colonizers. The leaders of the newly independent countries were far more hospitable to scientific expeditions than the European kings had been.

## Richard Spruce, Yorkshireman

Richard Spruce was born in 1817 in Gansthorpe, a village in Yorkshire, England. His mother died soon after he was born. Spruce's father, a schoolteacher, remarried a few years later. The new young wife gave birth to a succession of daughters and Dick's father and stepmother were busy with their growing family. Young Dick found himself free from parental supervision at a relatively early age. Fortunately, he was a bright boy and after completing school in the village he arranged for his own further education. For example, Spruce persuaded the local physician to tutor him in Latin and Greek.

On his own, Spruce gathered interesting plants from the moorland around the village and spent much time identifying and classifying his collection. By the age of 20, he had enlarged his collection to include 485 flowering species. Some of his rare finds were later included in catalogs of British plants.

Spruce followed in his father's footsteps and became a teacher at a college preparatory school. Although he did not teach botany, he spent his free time learning about that science. Spruce read many books on the subject and discussed his ideas with other amateur scientists.

Spruce also read accounts of South American expeditions and yearned to investigate plant life on that continent. After a year of teaching, Spruce realized that he did not enjoy the profession. However, he was virtually penniless and could not hope to find the resources to explore the New World. Spruce continued to teach and pursued his botanical studies as a sideline. Fortunately, his independent studies were carefully documented and his observations were accepted for publication in the scientific journals of the day. Spruce soon gained a good reputation among established scientists.

In 1844, Spruce's school went out of business and he needed to find another job. To further his interest in botany, Spruce had begun exchanging letters with Sir William Hooker. Hooker was associated with the British government's Botanical Gardens at Kew in suburban London. The older scientist unsuccessfully attempted to obtain a curatorship for Spruce at a museum in one or another of the British colonies. Hooker then introduced Spruce to George Bentham, a professional plant collector. Bentham had just returned from a successful plant-finding expedition in Spain and employed Spruce as an independent subcontractor. Bentham paid for Spruce to conduct an expedition to Spain. He also agreed to act as Spruce's agent in the sale of botanical specimens to museums and herbariums across Europe. The two-year enterprise was successful and they discussed an expedition to South America.

Two British naturalists, Alfred Wallace and Henry Bates, were already in South America surveying insects, birds, and animals. Spruce carefully read reports that the naturalists had sent back to England. He realized that his knowledge of botany would increase the depth of the investigations. The plans were completed for Spruce to join Wallace and Bates in South America. At last, he would realize his dream of exploring the new continent.

Spruce's ship departed from Liverpool in the early summer of 1849. It docked at Belém, Brazil, at the mouth of the Amazon River. As soon as he had rested from his trip, the young scientist began to collect exotic plants.

Although a few earlier Amazon explorers had some background in the study of botany, Spruce was the first to have the necessary knowledge, experience, and skills to undertake a thorough investigation of plant life in the area. After spending three months in Belém learning Portuguese and studying the local plants, Spruce embarked by sailboat for Santarém, about 500 miles (800 km) up the Amazon. The steady trade winds off the Atlantic pushed the boat against the river current.

Spruce collected plants whenever the boat came to shore. Along the way, he found a vine called sarsaparilla, which later became popular as a tonic and flavoring agent. Sarsaparilla is closely related to the Chinese plant that Rauwolf had found at Aleppo when he toured the Near East in the 1570s.

When Spruce reached Santarém, he met with Alfred Wallace and Henry Bates. Bates was in the process of collecting and identifying some 14,000 species of insects—8,000 of which were new to European scientists.

While staying in Santarém, the three English scientists experienced an unusually harsh rainy season. The downpour raised the water level by 40 feet (12 m) and the lands along the Amazon were flooded for hundreds of miles. The floods caused some dormant plants to bloom and Spruce was able to study these seldom-seen species. He spent a year in Santarém investigating the local plants.

Rubber trees were plentiful in the region and Spruce carefully recorded his observations of the harvesting of rubber. He noted that the native workers slashed the tree bark in several places and allowed the sap to flow into cups attached under the cuts. The sap of the tree was the raw material for making rubber.

From time to time, Spruce sent his botanical specimens back to Europe. His work was praised and more and more museums wanted his collections. By this time, Spruce had collected thousands of species and new ones were being added every day.

After Spruce left Santarém, he established a base at Munaus. His next venture was to travel up the Río Negro. During a stop in São Gabriel, a small village well upriver from Munaus, Spruce received a disturbing message. Wallace had contracted malaria and was near death. Spruce hastened to his friend's side and helped nurse him back to health.

After Wallace recovered, Spruce set out for the far upper reaches of the Río Uaupés. He again traveled by canoe and was forced to make frequent portages around the rapids and waterfalls. While on this journey, Spruce was introduced to a potent drug derived from a plant. During a religious ceremony, he

*Extracts from the roots of sarsaparilla have been used to treat many diseases and were once a popular ingredient in carbonated beverages.* (Courtesy of Alice Tangerini and the Smithsonian Institution)

drank a potion the natives called *caapi*. Fortunately, he did not have to drink a second cup. If so, Spruce would have experienced the hallucinations that the native Indians sought for themselves. Spruce recognized that this mind-bending plant material might have beneficial properties when used as a medicine. He reasoned that if the native population regarded the substance as potent, it was worth further investigation.

Spruce believed that native traditions were often valid and therefore important to his work. He studied the native languages and asked many questions about the local plants. Spruce recorded the native names, uses, and medicinal properties of the plants that he classified. His work was much more comprehensive than that of Aublet. Consequently, he is regarded by more scientists as the father of ethnobotany—the study of the relationship between native cultures and local plants.

Spruce sent the specialists at Kew Gardens samples and information about medicinal plants. Unfortunately, some cuttings—including the roots of the vine used to make *caapi*—spoiled before they reached England. A later sample of the vine was tested and revealed an alkaloid, a type of bitter enzyme that can induce hallucinations. Because enzymes have important medicinal properties, Spruce had been correct in his assessment of the plant. He listened ever more carefully when natives discussed the "magic powers" of local plants.

Spruce spent several more months collecting plants along the upper Orinoco and its tributaries. He was finally worn down by bouts of malaria, and in 1854 spent months near the town of San Fernando, Venezuela, recovering from the disease. When his health improved, Spruce headed back down the Río Negro to Brazil. When he reached Munaus, he found that the town was enjoying a booming economy in rubber. The price of rubber had increased by 500 percent over the past few years. Rubber was suddenly a major commodity on the world market.

Because of economic and political changes, Spruce's explorations now came under the direction of the British government. He was given specific orders to obtain cuttings and seeds of the cinchona tree, the source of quinine. To do this, Spruce had to travel westward to the headwaters of the Amazon and beyond. Unlike his other journeys, this trip began in high style. Spruce had obtained passage on a river steamboat. Although he traveled

*Richard Spruce recognized the value of plant knowledge held by shamans and became the first ethnobotanist.* (Courtesy of Kew Gardens, London, England)

in comfort, he was unhappy because the boat would not stop when he spotted a particularly attractive specimen along the shore.

The steamboat did not travel all the way to his destination and he continued by hiking through mountains and high valleys. This journey included occasional contact with a few Jivaro headhunters. The Jivaros and Spruce evidently developed a healthy mutual respect. The headhunters did not harm his party and provided them with food and drink.

Eventually, Spruce reached Ambato in Ecuador—a region where the cinchona tree flourishes. By this time, the collection of the bark was highly organized. Wealthy landowners controlled huge areas where cinchona trees grew wild. Spruce arranged to lease some of the land and collect the seeds as soon as they ripened. This venture required months to accomplish and needed an organized labor force to collect the ripe seeds before they fell to the ground.

Spruce made careful notes on the climate and soil conditions in which the trees seemed to do best. When plans were made to cultivate the cinchona tree in India, where malaria is a common disease, Spruce sent instructions to duplicate the growing conditions found in Ecuador. His advice was ignored by the British planters, and the cinchona plantations in India failed to thrive.

After fulfilling his duties to Queen Victoria, Spruce returned to Peru to continue his research. In 1864, after more bouts of malaria, he returned to England. The scientists at Kew Gardens invited Spruce to help organize the materials that he had sent back from South America. In addition, Spruce continued to work on his notes and revise the rough maps he had drawn during his expeditions.

Toward the end of his life, the British government gave Spruce a tiny pension and he retired to a cottage in Yorkshire. He died in 1893 before completing a summary of his experiences. His old friend Alfred Wallace edited his notes, and a two-volume account of his work was published in 1908. His recognition of the value of native knowledge was his key legacy. As they pursued their explorations, the scientists who entered the field after Spruce were able to save time and energy by consulting with native healers.

7

# The Story of Quinine

Malaria has been known to Europeans since the beginning of recorded history. Indeed, there is some archaeological evidence from human remains that malaria was a problem during Neolithic times. Until the modern era, the disease was widespread in the Mediterranean area and greatly feared throughout Europe.

Malaria is caused by a protozoan, a microbe larger than most bacteria. The microbe is transmitted by the bite of the female *Anopheles* mosquito. Standing water must be available for the mosquito to complete its life cycle. The female lays her egg masses in bodies of water that later support the hatchlings as they mature. Therefore, any wet, swampy area can be home to the *Anopheles* mosquito.

Europeans probably first learned of quinine in 1638. At that time, the wife of the Spanish viceroy (the royal governor), Countess Chinchón, was dying of malaria in Lima, Peru. In desperation, the royal physician recommended the use of a native cure—a powdered tree bark that the natives called *quinquina*. The powder was brought 500 miles (800 km) from the village of Loja in present-day Ecuador. The native treatment was successful. When the countess returned to Spain a few years later, she took along a supply of the powdered bark. The countess retained some of the medicine for herself, but she had

*The* Anopheles *mosquito is the carrier of malaria, one of the world's major killer diseases.* (Courtesy of the Office of Communications, U.S. Department of Agriculture)

another use for the rest. The poorly maintained family estate—about 25 miles (40 km) south of Madrid—was low-lying and badly drained. The peasant population was constantly sickened by malarial fevers. The powdered bark cured their fevers, and the workers soon restored the land to its former productivity.

There is no clear historical evidence that the story is true. Nevertheless, European botanists named the tree that supplies the medicinal bark after the Chinchón family. The medicine was called quinine, a European form of the native word quinquina.

South American Indians have used quinine to cure fever-producing illnesses for untold centuries. However, malaria was unknown to South America before the Spanish conquest in 1520. Spanish soldiers who already had the disease brought it to the New World.

Jesuit missionaries, sent to convert the natives to Christianity, were the first Europeans to see quinine as a potential treatment for malaria. The missionaries lived among the native peoples in remote areas where few Europeans dared to travel. The Jesuits probably observed the natives use quinine to treat many kinds of fevers. The Jesuits sent the powder back to Spain and to Rome, the headquarters of the Catholic Church.

At first, physicians were highly suspicious of quinine and the use of the drug was confined to the clergy. A high church official, Cardinal John de Lugo, promoted the use of quinine by both his malaria-ridden parishioners and his clergy. At that time, priests rather than doctors administered quinine.

In England, opposition came not only from the medical establishment but also from the strong Protestant majority. They disliked and distrusted the Catholic Church and could not tolerate the idea that anything good could come from the Jesuits. Strangely, the man who popularized the use of quinine in England was a self-proclaimed physician—a quack by the name of Robert Talbor.

Talbor had attended Cambridge University for a brief period of time. He had some training in pharmacy and had served as an apothecary's apprentice. Talbor, a bright but unscrupulous man, recognized that the bark was quite effective. He also recognized that it could not be sold under its popular name, Jesuit Powder. Talbor simply changed the name of the powder, masked the quinine's bitter taste by adding it to wine, and set up a pharmacy in Essex—a malaria-ridden area south of London. He advertised that he had a secret formula that did not contain Jesuit Powder. Talbor also advertised that Jesuit Powder had dangerous side effects that his medicine did not produce.

Because the "secret" formula worked well, and many malaria patients were relieved of their symptoms, Talbor's reputation spread rapidly. He traveled to London at the request of wealthy malaria victims and soon grew rich. The English king Charles II became one of his supporters, in spite of the opposition of the Royal College of Physicians. The king suffered from malaria and Talbor's remedy improved his health. In gratitude, Charles gave Talbor a knighthood. The quack became Sir Robert.

Louis XIV, the king of France, became his next royal patron. Both the king and his son were victims of malaria. Their recovery was hailed throughout Europe, and Sir Robert went on to repeat his successes in Vienna and Madrid. When he returned to Paris, he was hailed as a genius. King Louis gave him an aristocratic French name, and Sir Robert Talbor became Sir Robert Talbot.

When Sir Robert expressed the wish to return to England, Louis became upset. The king did not want the secret formula to slip away. Because quinine neither prevents nor cures malaria, the king was correct in thinking his family might require further doses. (Quinine helps suppress the multiplication of the malaria microbe and thereby prevents the severe fevers associated with the body's response to the microbe.)

King Louis proposed a deal. Sir Robert wrote the formula on a piece of paper, placed the paper in an envelope, and sealed the envelope. The formula was locked away—not to be opened until after Sir Robert died. In return, the king gave Sir Robert a handsome gift of money and a life pension.

Talbor died in 1681, shortly after his return to England. When the envelope was opened, the royal personages, wealthy patrons, and physicians were embarrassed to discover that the secret ingredient had been the hated Jesuit Powder. The crafty Talbor had suspended the powdered bark of the cinchona tree in white wine where it made a bitter but acceptable elixir. People began to revise their attitude toward Jesuit Powder.

## More Chemistry

In 1820, two young Frenchmen working in Paris. Joseph Pelletier and Joseph Caventou succeeded in isolating the key ingredient in quinine. They found it to be a very complicated enzyme molecule called an alkaloid. This molecule is composed of three carbon rings. Two of the rings have one atom of nitrogen and two of carbon as attachments. The third ring is unusual because the attached carbon atoms connect to a structure with an oxygen and hydrogen atom at its end.

The identification of the curative ingredient in quinine was important for three reasons. First, raw bark could be tested for its quinine content, and payment could be based on the amount of that ingredient. Second, because the active ingredient could be weighed, doctors could prescribe the exact amount needed for proper treatment. The tendency to overdose patients was

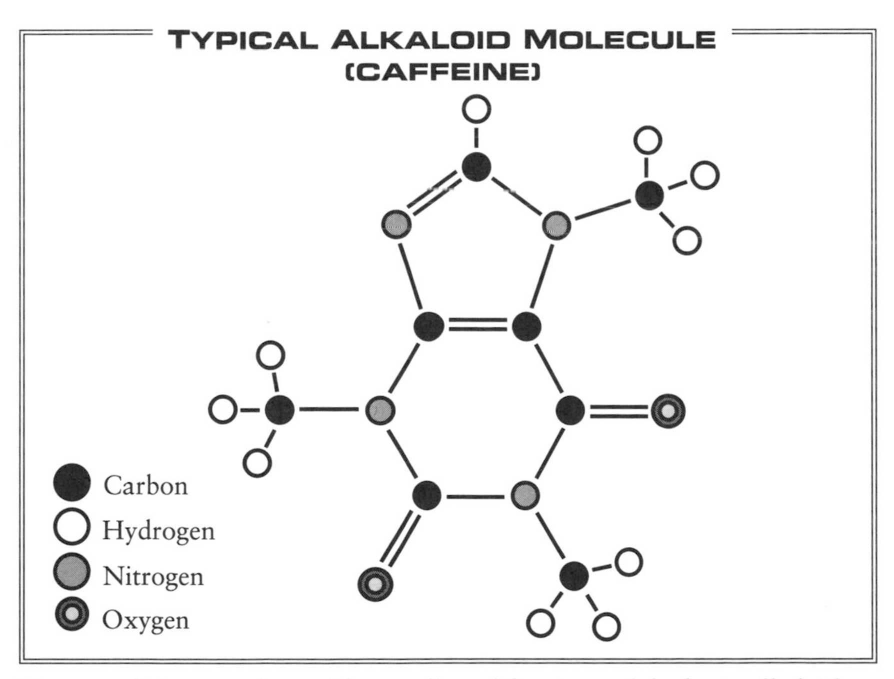


*Many medicines are formed by small modifications of the basic alkaloid molecule.*

greatly reduced. The third advantage concerned the new, easier way in which quinine could be administered. The pure alkaloid could be extracted from the raw bark and formed into small pills. The bark's bitter taste was no longer a problem.

## Breaking the Monopoly

The first half of the 1800s was a time of political unrest in much of Latin America. The people were fighting for their independence from European rule, and governmental authority was often lacking. Bandit groups and armed militias controlled many parts of the countryside, and this confusion interfered with the availability of quinine.

Major land owners controlled the harvesting of the wild cinchona trees and the world market depended on such supplies. The number of wild trees was rapidly decreasing because stripping away the quinine-rich bark killed the tree. Also, the transportation routes from the forests to the ports were often interrupted by local warfare. The price of cinchona bark rose as it became a scarce commodity.

The Dutch had grave problems with malaria among the colonists and native workers in the Dutch-owned East Indies, present-day Indonesia. They hoped to solve that problem and make a handsome profit by cultivating cinchona trees on Java, an island in their colony.

The Dutch government sent a botanist, J. C. Hasskarl, to collect seeds and seedlings in Peru and Bolivia. Dutch diplomats in Peru were told to make similar collections. The plant material was placed on a Dutch warship headed for present-day Djakarta, the capital of Indonesia. The voyage took a long time and few seeds or seedlings survived. Those that did produced bark with a low quinine content and the project was an economic failure.

In 1859, Clement Markham, an Englishman, persuaded his government to develop plantations of cinchona trees in India. Richard Spruce, the noted South American explorer and bota-

*The bark of the cinchona tree is the source of quinine, an alkaloid drug used to treat malaria.* (Courtesy of Alice Tangerini and the Smithsonian Institution)

nist, was assigned to collect the seeds. Although a million trees were planned for southern India and Ceylon (now Sri Lanka), the planting was poorly organized and the trees did not prosper.

Charles Ledger, a British citizen who lived in Bolivia, was the next to threaten the South American monopoly on quinine. Ledger was a professional bark dealer who employed natives to locate wild cinchona trees and harvest their bark. He assigned one of the workers to collect seeds from very productive trees. Ledger hoped to sell the seeds to the British government and thus reawaken their interest in cultivating cinchona trees. Ledger sent 14 pounds (6.3 kg) of seeds to his brother, George, who offered them for sale. The government officials had bad memories about Clement Markham's plan and declined to do business with the Ledger brothers.

George Ledger then asked the Dutch if they were interested in the seeds. Their officials were cautious but agreed to purchase a pound at a very low price. The other 13 pounds (5.9 kg) were sold to a British planter on his way to India. The planter traded them for different seeds as soon as he arrived at his destination. The 13 pounds (5.9 kg) of cinchona seeds were mishandled by the seed traders and never germinated.

The pound of seeds purchased by the Dutch was handled far more carefully. The cinchona seeds were shipped to Java, where growing conditions matched those in Peru and Bolivia. The quinine content of the healthy young plants tested three to four times higher than those from the prior plantings. The Dutch, through their greater botanical skill, had broken the South American monopoly and now controlled the most productive source of quinine.

The unfortunate native who had gathered the seeds in Bolivia was thrown into jail for violating the Bolivian prohibition against collecting cinchona seeds. Undoubtedly, the poor man was unaware that he was responsible for assuring a steady supply of one of the world's most important medicines.

Atabrine, the first synthetic medicine for malaria, was not developed until 1926. This synthetic drug helped break the Dutch dominance of the quinine market.

# 8

# The Story of Curare

The interior of South America was avoided by early colonists because of its difficult terrain. Consequently, many of the original inhabitants were able to continue their tribal practices with few modifications. Indeed, the peoples who lived on the eastern slopes of the Andes Mountains and in the rain forests were almost totally free from European interference. Their villages were populated by extended family groups and sustained by some farming, hunting, fishing, and the gathering of wild fruits and vegetables. Hunters used both bows with arrows and blowguns with small darts. Often, the tips of the darts and arrows were coated with a dark brown poisonous paste. This poison has many names but is known in the English-speaking world as curare.

## Initial Explorations

Spaniards and other Europeans were intrigued by curare from the beginning of colonization. The Spanish explorer Francisco de Orellana wrote about the effects of curare in 1541. He noted that when the poison enters the bloodstream of a bird or animal it generates a general paralysis that is followed by death.

The Spanish were unsuccessful in their attempts to learn more about this strange substance. Because the native peoples had often been treated badly by the Spanish, it is not surprising that they refused to share their recipes for making curare. Over the years, a few explorers managed to obtain samples of the dark brown paste, but no one could obtain the recipe and no one could isolate the ingredients. Though the effects of curare were little understood, scientists speculated that this potent poison might have a possible use in medicine. They knew that curare induces a total relaxation of all muscles except the heart. Therefore, some reasoned that the substance might be used to treat diseases such as epilepsy that cause dangerous muscle spasms.

## Setting the Stage

An Englishman named Charles Waterton was one of the first to actively pursue the idea that curare could be used in a medicine. Waterton had inherited large farming estates in Yorkshire, England, and was responsible for the family's sugar plantations in present-day Guyana, a small country on the northeastern coast of South America. His plantations were on the Demerara River near Georgetown, the capital of the country.

In 1810, while Waterton was managing the family's Guyanan plantations, he had an encounter with the local police. They were seeking a friend of Waterton's in connection with some bad debts. Waterton refused to help the police in their quest and was soon called before the governor-general of the colony. When Waterton did not deny his actions, the governor-general was impressed by his courage and released him. The official later became Waterton's friend and awarded him a permit to explore the upper reaches of the Demerara River.

Waterton, a widower in his late 20s, was an adventurous person. He was a notable hunter and amateur naturalist. He was also a field biologist with a great interest in birds. Waterton saw his trip up the Demerara River and into the interior as a major opportunity to search for new species of birds.

As a hunter, Waterton was aware of the use of curare by the native peoples and was curious about its properties. He thought that curare might provide a cure for tetanus, rabies, and other diseases that induce powerful muscle spasms. Waterton also had ideas about how accidental curare poisoning might be remedied.

Perhaps because the natives he encountered had had no previous contact with Europeans, the young adventurer learned more about curare poison than any previous explorer. Waterton was given samples of the material and shown how it was prepared. He learned that the main ingredient was obtained by stewing pieces of a certain woody vine in boiling water. Other plant materials were then added to make it gluey, and ingredients such as ants and pepper pods were added for dramatic effect.

Waterton took some of the curare back to England. After he returned to his home in Yorkshire, he tested one of his ideas about the effects of the poison. Waterton demonstrated that his donkey could be kept alive by artificial respiration after being injected with the poison. His crude—but successful—artificial respiration technique involved pumping air into the animal's lungs by using a bellows borrowed from the family fireplace. He proved that the effect of the poison was short-lived and that a victim could survive if administered oxygen during the time of paralysis. However, the short duration of the poison's effect meant that chronic diseases with persistent muscle spasms were not helped by curare. Repeated doses—to extend the period of relief—were too dangerous for the patient. Waterton's donkey was unwell for about one year after her experience with curare.

## Solving the Curare Puzzle

Although curare was not useful as a cure for chronic muscle seizures, some members of the medical profession continued to show an interest in the substance. Over the following century, doctors and chemists in Europe and South America requested samples for testing their ideas. Chemists kept trying, unsuccessfully, to isolate the active ingredients. Some physicians used small amounts of the material to moderate the cramped muscles caused by severe arthritis. However, most doctors were afraid to prescribe the poorly understood compound.

In the 1930s, interest in curare began to surface again. About that same time, Richard Gill, a young American from Washington, D.C., became curious about the substance. Gill and his wife, Ruth, owned and operated a ranch in Ecuador. The ranch was located in the valley of the Río Pastaza, about a day's horseback ride from the nearest town. Although somewhat isolated, the ranch was on a major thoroughfare for natives traveling between the eastern slope of the Andes and the tropical jungles that grew in the plain. Gill came to know all the frequent travelers. He also became friends with the local Quichua Indians who worked for him, and he learned their language.

Gill's interest in native medicines was aroused when Ruth had a bad fall. Gill was preparing to carry her by stretcher into the nearest town when one of the ranch hands declared that the long, difficult trek might not be necessary. The elderly Indian said he had studied with local shamans and had traveled, in his youth, to the territory of the feared Jivaro tribe, noted for its drug-making skills. Gill was persuaded to let the Indian treat his wife in the local manner. When Ruth recovered after the treatment, Gill was convinced that there were healing powers in native medicines.

Years later, when Gill was back in the United States on a vacation, he experienced symptoms of multiple sclerosis (MS), a degenerative disease of the central nervous system. An attending physician mentioned that curare might provide at least some release from the crippling pain and tremors of MS. The doctor also remarked that the identity of the active ingredient in curare was unknown and dosage control was difficult. When asked, the doctor told Gill that the supply of the raw plant materials was insufficient for an extensive chemical analysis of curare.

Gill resolved to overcome his physical disability, return to Ecuador, and learn the secrets of curare from the Indians. He was confident that he could discover which plants were used to make the substance and the methods to manufacture it.

While Gill convalesced, he wrote popular articles about his interest in curare and his adventures as an Ecuadorian rancher. In 1938, Sayre Merrill, a wealthy New England businessman, read one of his stories. Merrill offered to sponsor an expedition with Gill as leader. After careful planning and preparation, Gill and his wife returned to Ecuador.

Gill had been gone for six years and his property was not in good condition. He rehired his farmhands and quickly restored the ranch. Using his home as a base, Gill recruited nearly 100 local men. He also purchased or hired mules and horses to carry supplies and trading goods into the interior. After everything was organized, they headed east into a jungle region called the Pacayacu. Because of his disability, Gill had some difficulty negotiating jungle trails in the Andean foothills and the progress was a slow. After a few days, they reached one of the tributaries of the Amazon River, where they traveled faster and more comfortably by dugout canoes. In order to preserve the secrecy of his sources, Gill never revealed which tributary he embarked upon.

Gill sought a major village of the Jivaro people. These natives were noted for their use of curare and their resistance to outside influences. In spite of their reputation for hostility, Gill gained the confidence of the village leaders. Ultimately, he was allowed to observe the full ritual of curare preparation. Gill learned that the villagers used as many as six different plants in the preparation. Three plants were used in the first stage of brewing and the bark of a fourth was added after the mixture had boiled for a while. Two days later, two more plants were added and the stew was slowly boiled down to a dark brown paste.

After comparing several different recipes, Gill discovered that the one common ingredient was a vinelike plant with a stem that measured 3 to 5 inches (8 to 13 cm) in diameter. As Waterton had reported about one hundred years before, the vine was cut into short lengths and stewed to extract the critical ingredient.

When Gill and his wife left the jungle after their three-month adventure, they carried samples of the plant and the semiprocessed poison. They returned to New York City, where Gill wanted to start a business trading in curare and the raw materials from which curare is extracted. Gill expected to retain his monopoly on the curare market because no one else had gained the cooperation of the Jivaro people.

At first, as in Waterton's time, the medical community was enthusiastic about the possible uses of curare. Some specialists hoped that it might help in the treatment of epilepsy and other diseases. The major pharmaceutical houses, however, soon lost interest. Scientists rediscovered the fact that the drug's actions were too transient. After receiving a dose of the medicine, a patient's muscle spasms decreased but the drug's effect soon wore off. Curare was classified as a palliative—a medicine that reduces the pain or intensity of a medical condition.

A few years later, however, the drug was proven to be invaluable in the medication of surgery patients. Surgeons needed a way to relax muscles—particularly of the abdomen—during operations. Ordinary anesthesia might put the patient to sleep and deaden the pain sensations but did little to relax tensed muscles. Curare worked well for this purpose. Its short-lived effects were ideal because physicians wanted muscle tone to return as soon as the operation was completed.

In spite of this important development, Gill never became a major trader in curare or its raw materials. The large pharmaceutical houses developed their own suppliers and immediately broke Gill's monopoly. He tried unsuccessfully to find a backer to finance another, grander expedition into the Amazon region.

The National Research Council proposed a government-funded project. Dr. Robert Griggs was chosen to be leader and Gill was offered the position of field assistant. Because Gill was not a trained scientist, Griggs thought that Gill could contribute little to a scientific expedition. Therefore, Gill was assigned to guide the group and make contact with the local tribes. He was insulted and disappointed and presented his grievance to a government panel. No verdict was reached because the National Research Council withdrew their proposal for the expedition.

Gill moved to California and supported himself by importing and reselling curare paste and other tropical products. In his garage laboratory, he worked on new ways to employ curare. Richard Gill died in 1958 of complications of the disease that he had acquired 20 years before. Through his efforts, curare became an important medical asset. Curare's greatest contribution to health care is the use of its active ingredients as chemical building blocks. One of the active agents was found to be a perfect catalyst in the synthesis of a fast-acting anesthetic.

In modern times, the mystery of the raw materials and chemical composition of curare has been solved. Field botanists and anthropologists learned that at least seven different plants can be used as the principal raw material for making curare. Different native groups use different combinations of raw materials as well as different incidental ingredients. Consequently, there may be hundreds of recipes for curare and probably no two batches are exactly the same—even if prepared by the same shaman.

The identification of the key active ingredient was difficult and time consuming. At first, chemists were mystified when each new sample of raw curare yielded a totally different group of components. They expected to find one common active ingredient. Although the seven main plant materials all cause rapid general muscle relaxation, the active ingredients of each material have slightly different chemical compositions. The chemists finally realized that they were dealing with not one but several highly similar alkaloid molecules that produced the same physical effects.

9

# Hormones

Some molecules manufactured by the human body are closely paralleled by some molecules produced by plants. In many cases, a plant-derived molecule can block the actions of a chemical produced by the human body. When morphine, heroin, or other drugs derived from the opium poppy are introduced into an individual, their molecules block chemicals that help keep the person alert and sensitive to pain. Consequently, the person becomes drowsy and anesthetized.

In addition to blocking the effects of some of the body's natural chemicals, plant molecules can substitute for the catalysts in the human body. Catalysts are essential to life because they activate vital chemical reactions. Enzymes are catalysts that cause molecules to form necessary compounds. Hormones—from a Greek word that means "urge on"—are also catalysts but serve a different function. A hormone is produced by one human organ and then carried by the bloodstream to a different organ. As the Greek word implies, the hormone then "urges on," or stimulates—by acting as a catalyst—that second organ to produce certain chemicals. Serious diseases such as diabetes result when the body fails to manufacture enough of a particular hormone. Over the past 150 years, scientists have identified all or nearly all of the vital hormones. Fortunately, hormone substitutes made from plants or synthetic materials can act as exact copies of many of the missing hormones and allow a human body to function properly.

Human organs known as endocrine glands produce dozens of hormones that support various body functions such as storing energy and activating muscles. In some ways, these hormones are similar to vitamins in that only a small quantity is needed at any one time—but that small quantity is very important. The hormones control various nonglandular organs and allow one hormone to regulate the amount of another hormone in the system. Scientists have thoroughly investigated the hormonal activity of the adrenal gland and the reproductive organs and have searched for plant-derived or synthetic hormones to mimic the natural effects. Over the years, they have arrived at some far-reaching conclusions.

## The Adrenal Glands

The interior of the adrenal glands produces two hormones of major significance—epinephrine and nor-epinephrine. (These hormones are also known as adrenaline and nor-adrenaline.) They are released when the body senses a severe physical threat. The brain recognizes the threat and the nervous system triggers these hormones. Indeed, there is a strong link between these glands and the nervous system. Epinephrine and nor-epinephrine activate the "fight or flight" reflexes. In other words, when the two hormones are released, the human body is ready to fight a "battle" or run the other way. The nervous system acts to draw fuel from the liver, increase the heart rate and blood pressure, and—in general—pep up the body for intense activity.

The significance of the adrenal glands was discovered by Dr. Thomas Addison, who was born in Scotland and studied medicine there. In 1855, when Addison was on the surgical staff of Guy's Hospital in London, England, he noted a mysterious, often fatal illness. The patients suffered from a general weakness,

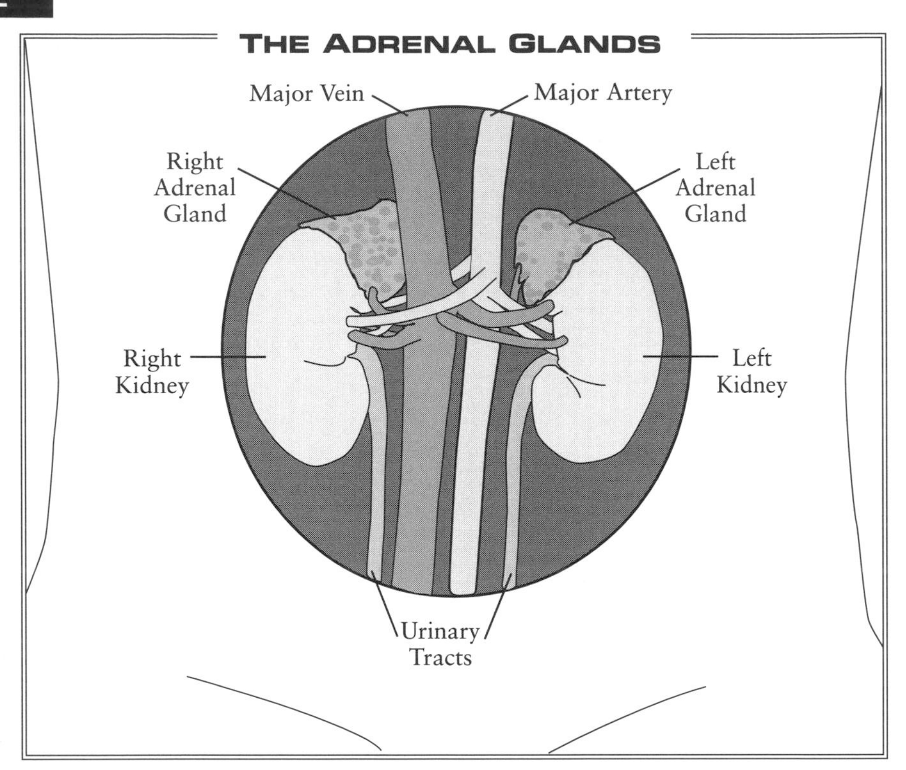


*The adrenal glands are sources of many hormones; they sit atop the kidneys.*

anemia, and a yellowing of the skin. Many cases were so alike that Dr. Addison correctly concluded that he was seeing the effects of one particular disease. After the patients died, he conducted detailed autopsies of the bodies and found only one factor in common: each autopsy showed that the small oval organs that lay over each kidney were severely shriveled. Addison concluded that the disease was caused by the malfunction of these organs—the adrenal glands. The condition has been named Addison's Disease after the man who identified the cause.

Later, an American of French ancestry named Charles Brown-Sequard showed that animals do not survive after the removal of their adrenal glands. In short, these seemingly insignificant organs are essential to life. Unfortunately, neither Addison nor Brown-Sequard were able to determine why these organs were so important.

# An Adrenaline Mimic

Thousands of years ago, people in China brewed a tea made from a wiry-stemmed plant called *ma huang* in the Chinese language. The modern botanical name for the plant is *Ephedra sinica*. The tea was a famous stimulant and used in ancient times for the treatment of hay fever and related breathing problems. Teas made from closely related plants were drunk over the course of time by many peoples in many places. Indeed, tea made from a close relative of *Ephedra sinica* was brewed by the Greeks and Romans. In A.D. 60, the Roman naturalist Pliny the Elder wrote a description of the tea's effects.

A similar plant was discovered by the pioneers who built the American southwest. They used another member of the Ephedra family for their stimulating tea. Thus, for centuries, some form of Ephedra tea was a common folk remedy although no one understood how or why it worked. As usual, there are many

*Stems and leaves of the female ephedra plant shown here have been used to make a stimulating tea by many different people in different parts of the world.* (Courtesy of James Manhart, Texas A&M University)

stories—both true and false—about popular folk remedies. Ephedra tea was once thought to be a cure for syphilis. In reality, the plant had no curative effect on that disease.

In 1885, the active ingredient in ma haung was isolated by Japanese scientists. The ingredient, an alkaloid, was purified in 1887 and named ephedrine. Around 1924, an American and a Chinese-American physiologist conducted clinical tests on the compound while working at Peking Union Medical College. The two men had held a long-standing interest in Chinese folk medicine. Their introduction of ephedrine into standard medical procedure was a major success. For example, it is used to treat the symptoms of asthma and as a heart stimulant in cases of acute heart stoppage.

Chemists soon discovered that ephedrine—the plant product —has a chemical makeup and properties similar to epinephrine—the human product. The two chemicals are both alkaloids and share the same basic molecular structure. The plant-derived medication, however, has some advantages. Ephedrine can be taken by mouth while epinephrine loses its effectiveness in the digestive system and must be administered by injection. For some conditions, such as asthma, ephedrine is the better choice. Both the animal hormone and the plant hormone can be used in the treatment of several conditions including sudden heart failure. If the heart stops beating because of a severe injury or an overdose of drugs, paramedics usually administer epinephrine or ephedrine along with electrical stimulation to restart a heart beat.

Once the chemical connection between epinephrine and ephedrine was made clear, chemists began to study similar alkaloids. They added and subtracted combinations of atoms that were connected to the basic alkaloid structure of epinephrine. This work soon led to the development of benzedrine and a growing family of amphetamines—called pep pills—all derived from the plant material, ephedrine. The "rush" some people experience when riding a roller coaster is imitated by benzedrine and the other amphetamines. This effect leads to some misuse of these drugs.

## The Adrenal Cortex

When the adrenal glands were found to be the source of powerful hormones such as epinephrine (adrenaline), scientists began to search for other hormones. In 1924, researchers at Johns Hopkins Hospital in Baltimore, Maryland, proved that Addison's Disease is not caused by a shortage of epinephrine—as Addison had believed. Instead, the illness is caused by a lack of material from the cortex (outer skin) of the adrenal gland. Studies found that the cortex of the adrenal glands produces dozens of complicated chemical compounds. The compounds counteract inflammations, allergies, and the negative effects of stress and are more important to the human body than adrenaline. In 1935, a pure extract of the outer skin of the adrenal glands was produced by scientists at the Upjohn Company. By 1940, 40 of the 46 distinct compounds in this extract had been isolated and identified.

## The Search for Cortisone

Today, cortisone is produced in large quantities and at low cost from plants such as the Mexican yam. However, it was very rare until the 1960s. In late 1941, just after the United States had entered World War II, a strange rumor circulated among the military services. The rumor indicated that German bomber pilots were being given an extract from the adrenal cortex to counter the effects of high-altitude flights. This rumor was false but it mobilized the U.S. research establishment. Scientists soon realized that every cow and pig in the country would lose their adrenal glands if the government ordered enough extract to treat all U.S. bomber pilots. That project was not too practical and a synthetic substitute was sought.

The first attempts to find a substitute for the extract from the adrenal cortex used a compound found in the bile of cattle—a waste product of the beef industry. This compound is closely related to cholesterol. Researchers at Merck and Company, the drug manufacturer, used as many as 60 steps to complete the synthesis. The scientists synthesized a few grams of three different chemicals produced naturally by the adrenal cortex. By the time this work was completed, World War II was over. Most of the chemicals were used to treat victims of Addison's disease, but a few grams were put into storage in the laboratory.

In 1948, Dr. Philip S. Hench at the Mayo Clinic in Rochester, Minnesota, observed that some female patients with severe arthritis showed great improvement if they became pregnant. Some patients with jaundice showed the same effect. At first, the connection between pregnancy and the relief of symptoms of arthritis or jaundice was not clear. In searching for a link, researchers surmised that the improvement might be due to a stress reaction. Pregnancy or jaundice might be a source of stress that activated the adrenal glands. The glands, in turn, might produce a hormone that reduced the swelling and inflammation of arthritis or improved the condition of a jaundiced, pregnant woman.

When a patient with severe symptoms of arthritis was admitted to the clinic, Hench decided to test his ideas. He obtained a small sample of the chemicals that had been produced by the Merck scientists. Using a fraction of the compound, he made a solution and injected it into the vein of the patient. After two days of treatment, the patient's pain, stiffness, and swollen joints were gone. The material, which had been labeled Kendall's Compound E, was soon renamed cortisone.

Medical applications of cortisone grew rapidly, but the supply remained tiny. The only commercial source was an extract made from the bile removed from slaughtered cattle. No major drug firm had been able to invent a new way to mass produce cortisone.

## The Testes

In the meantime, other human glands were being investigated and scientists were reading earlier studies on the subject. In

1849, the German physiologist Arnold Berthold demonstrated that masculine behavior could be restored to a castrated chicken when testes from another male chicken were surgically implanted. This research showed that male sex organs did more than produce sperm. The "foreign" testes contributed some substance to the rooster's body that activated male patterns of behavior. However, at that time, no one could identify the substance. In 1927, almost 75 years later, a chemical was extracted from bulls' testes and administered to castrated roosters. The chemical restored the roosters' masculine behavior. Because the material was effective in a different species of animal, the scientists, F. C. Koch and L. C. McGee, proved that the chemical extracted from the testes was a universal restorative of masculine characteristics.

The work of Koch and McGee was published in 1927 but another eight years were needed to identify the chemical structure of the substance. The masculinity-inducing hormone was actually two slightly different chemicals—androsterone and testosterone. They belong to the chemical family of steroids, fatlike materials that include cholesterol. Biochemists have recently shown that both male and female hormones use cholesterol as a raw material.

When a man's body cannot produce a sufficient amount of testosterone, an injection of the male hormone can restore masculine characteristics and functions. The same hormone has the ability to activate muscle growth. After stomach surgery, patients often lose their appetites and suffer from malnutrition. They cannot manufacture sufficient protein to build muscle tissue. Indeed, their muscles slowly deteriorate as muscle cells die and are not replaced.

Injections of male hormones restore the patient's appetite and the ability to manufacture and use proteins. Weight loss is stopped and muscles begin to grow. The use of male hormones is of special interest to doctors who treat young children for nervous disorders and failure to develop good musculature. Unfortunately, the hormone treatments that build bodies also bring about secondary sex characteristics such as facial hair.

Chemists began to seek synthetics that would have the positive effects of the natural hormone but would not masculinize young children and female patients. Such synthetics were achieved by scientists at the Searle Company in 1955.

These synthetic male hormones are particularly effective in treating burn victims. When young children are badly burned, the hormones are especially helpful in healing large blisters and speeding recovery.

## The Ovaries

Almost 50 years after the experiment with a transplanted rooster testes revealed the function of the male hormone, a similar experiment was conducted by transplanting ovaries. The studies were done in Austria in 1896. However, the effects of ovarian extracts were not demonstrated until 1923. Several more years were needed to isolate the active ingredient. This was finally achieved during the search for a quick and easy pregnancy test. Two German scientists found that if the urine of a pregnant woman was injected into an immature rabbit, the rabbit would become sexually receptive. The new test seemed to show that a pregnant woman's urine contained a fair amount of the female hormone. So, chemists began to analyze samples of the urine to isolate the hormone. They soon discovered that the urine of pregnant female horses also contained the female hormone and was more easily obtainable. Therefore, horse urine became the raw material of choice. Pure crystals of the female hormone, estrone, were isolated in 1929. Shortly after, two other female hormones were identified and isolated—estrogen and progesterone.

The main application for these female hormones was to regularize the menstrual cycle, which can be upset by illness, injury, or other sources of stress. However, in the 1940s and early 1950s, the most important use for estrogen was to help childless women become pregnant. Through the administration of rather large doses of the hormone, patients were induced to experience a "false" pregnancy. For several months, symptoms such as morning sickness were artificially induced. Scientists hoped that the body of the potential mother would initiate pregnancy-like conditions in the womb and other organs. In other words, the body would have a rehearsal for being pregnant. When the hormone injections were stopped, the body was better prepared for a real pregnancy. This scheme worked for a large proportion of infertile patients.

Doctors found that during a false pregnancy—induced by the hormone injections—and a real pregnancy women stopped producing eggs. This discovery instigated the development of the contraceptive pill—a medication that prevents women from becoming pregnant.

Up until the late 1940s and early 1950s, all the sex hormones and cortisone, the close chemical relative of the sex hormone, were in short supply. These materials were all derived from animal sources. However, the supply problem was solved by the efforts of several chemists who discovered that these hormones could be produced from plant extracts in large amounts at low cost. This change of raw material source has had and continues to have major societal effects in the areas of disease prevention and cure and in the control of human reproduction.

# 10

# Synthesis

In 1940, an important historical document was brought to light by Dr. E. W. Emmart, a professor of medicine from Johns Hopkins University in Baltimore, Maryland. While visiting the Vatican Library in Rome, Dr. Emmart, an ardent Catholic and book lover, came across a little-known work called the Badianus Manuscript. In the early 1550s, the author, an Aztec physician, had compiled a complete list of all the medicinal plants used in traditional Aztec medicines. The document, which he wrote while teaching in central Mexico, contained descriptions of more than 250 plants. The manuscripts included information about whether a particular plant was cultivated or grew wild. It also told how the medicinal ingredients were extracted and how the medicines were used in the treatment of various diseases.

In 1552, the book was translated from the Aztec language into Latin by a native named Juan Badiano and the translation was then sent to Spain. The history of its journey to the Vatican Library is unknown. Dr. Emmart's discovery of that 400-year-old manuscript generated new public and professional interest in natural products and folk medicine.

## Russell Marker

One of the plants mentioned in the Aztec herbal was the Central American yam. This plant is not to be confused with the sweet potato that is sometimes called a yam in North America. Some Central American yams are edible but most of the 400 species are either unpalatable or poisonous. In fact, a fish poison is made from the mashed root of certain species. When the substance is released into a lake or river, the fish in the immediate area are killed quickly, but their flesh remains wholesome. Other species of yams produce a juice that foams like soap and the natives use it for washing clothing.

In 1940, the same year that the Badianus Manuscript stimulated an interest in medicinal plants, a biochemist named Russell E. Marker was working on a new line of research. For some time, he had been thinking about synthesizing hormones—particularly female sex hormones such as estrogen and progesterone.

Scientists knew that the symptoms of rheumatoid arthritis in a female patient were significantly relieved if she became pregnant. There was a strong theory that the pregnancy-related increase in progesterone—a female hormone—was the key factor in reducing the symptoms. The sex hormones and cortisone are members of the family of chemicals known as steroids. The steroid molecules are similar to cholesterol and are composed of four rings made up of carbon, oxygen, and hydrogen atoms.

In the early 1940s, male and female hormones and cortisone were difficult to obtain and extremely expensive. At the same time, scientists knew that hundreds of natural steroids were readily available from a variety of plants including the Mexican yam. Marker thought that perhaps a species of yam might supply the perfect steroid to facilitate the synthesis of hormones—particularly progesterone.

Marker had been studying natural steroids for years. He was on the chemistry faculty at Pennsylvania State University and worked part-time for a major pharmaceutical house. Over the years, he had written more than 150 research reports and had received more than 75 patents on steroid chemistry.

In addition to his other talents, Marker had learned to be a competent field botanist by spending his vacations in Mexico, making surveys of growing plants, and gathering information about the properties of local plants—particularly the yams.

Marker was most interested in the steroid diosgenin closely related to sapogenin, used by villagers as soap. He discovered that the best source of diosgenin was a yam that grew from a rough-skinned, dark brown rhizome or potato-like root. The root usually weighs about 4 pounds (2 kg), but older plants have larger roots that can weigh more than 100 pounds (45 kg). Marker had found the ideal raw material for his experiments.

In 1943, Marker resigned from his professional positions and moved to Mexico. Just outside Mexico City, he rented a modest shed to use as his chemistry laboratory. Using a secret process, he produced over 4 pounds (2 kg) of pure progesterone in a few

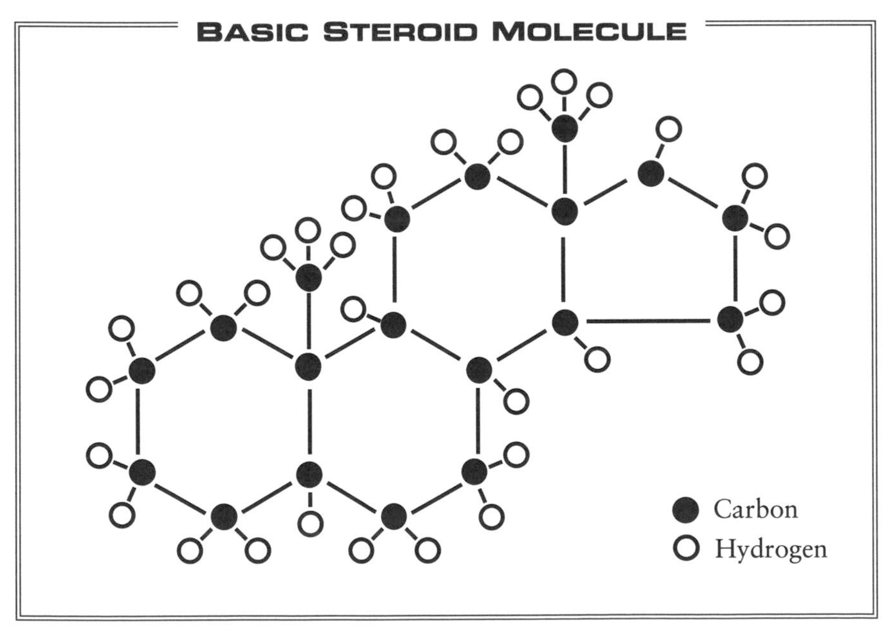


*The steroid molecule is the building block for cholesterol and several hormones such as cortisone and the gender hormones.*

*The tuberous roots of the Mexican yam, often weighing as much as 55 pounds (25 kg), are the source of a steroid that can be converted into several hormones.* (Courtesy of James Manhart, Texas A&M University)

months. At the time, progesterone's wholesale price was about $40,000 a pound. Marker sold all 4 pounds (2 kg) to Laboratorios Hormona, a Mexican company established 10 years before by two European refugees. The owners purchased Marker's homemade progesterone and offered him a partnership in Syntex, their newly formed company. Syntex broke the European virtual monopoly on progesterone and reduced its price by 90 percent.

At first, all went well with the partnership. Soon, however, it became apparent that Marker was a difficult person and did not relate well with the others. The partners bought out Marker's share of the business in 1945. Once again, Marker worked on his own and after some initial success dropped out of public view.

Meanwhile, the founders of Syntex hired another European refugee, George Rosenkrantz, a Swiss-educated Hungarian. In 1951, Rosenkrantz hired Karl Djerassi, a young expatriate

Bulgarian, already well regarded as a steroid chemist. Djerassi and Rosenkrantz immediately began research on a process to synthesize cortisone from yams.

## Karl Djerassi

Djerassi was born in 1923 and lived in Sofia, Bulgaria, until his parents divorced. After the divorce, when Karl was six years old, his mother, an oral surgeon, returned to her home city of Vienna, Austria. Until the beginnings of World War II, Karl spent the school year in Vienna with his mother and summers in Bulgaria with his father. His father, also a physician, was a specialist in chronic diseases.

In 1939, when Hitler took over Austria, Karl's parents remarried so that the boy and his mother could claim Bulgarian citizenship. Their new passports allowed them to escape first to Sofia and then to the United States. After arriving in New York, they were helped by the Hebrew Sheltering and Immigrant Aid Society. Karl's mother found a job as a medical assistant in rural New York State.

Karl, who had had only two years of secondary education in Europe, decided to begin his higher education as soon as possible. He wanted to become a physician. In December, Karl went to see a friend of his father's who was a faculty member at New York University (NYU). Karl was advised to apply to Newark Junior College in New Jersey since NYU did not accept students in the middle of a school year. His transcript from the American College of Sofia—a school that he attended briefly—persuaded the officials at the New Jersey college that Karl had had at least one year of higher education. He was accepted at the junior college.

He realized quickly that the Newark Junior College was not a good springboard for a career in medicine. However, he took the premedical science curriculum and was inspired by the chemistry teacher to consider chemistry as a career choice. In the meantime, he applied for a foreign student scholarship from the Institute of International Education. He obtained the scholarship and was accepted at a small four-year institution, Tarkio College in Tarkio, Missouri.

Through a series of strange coincidences, Karl became a frequent speaker at church suppers. He always spoke on the "current situation in Europe." Perhaps because of his foreign accent, he was well received. Actually, he knew little about the subject and got most of his ideas from John Gunther's book, *Inside Europe*. The donations given by audiences at the churches or luncheon clubs kept him in pocket money while he was at Tarkio.

Djerassi learned that he was being considered for a scholarship at Kenyon College in Gambier, Ohio. That June, he stopped to visit the school on the way to his mother's home in upstate New York. He fell in love with the Episcopalian institution and was delighted to accept the scholarship. Kenyon had an outstanding chemistry faculty, and Djerassi soon decided to change his major from premed to chemistry.

Djerassi was 18 when he finished his degree. He had attended Kenyon for two semesters (with extra-heavy course loads) and a summer term. Djerassi was not called into the military service because of an old knee injury. Therefore, he was free to find a job or continue his education.

After graduation, he went to visit his mother in Ellenburg Center, New York. While there, he read some publicity material from pharmaceutical companies and realized that many of them had production facilities in nearby New Jersey. He sent letters of application to the research departments of several of the companies. Most did not answer his letters, but he was invited for an interview by the CIBA Corporation.

Djerassi was hired and assigned to be an assistant to Charles Hutterer—a fellow refugee from Vienna. Hutterer soon succeeded in synthesizing pyribenzamine, one of the first effective antihistamines—a medication that relieves the symptoms of colds and allergies. Djerassi was included in the list of inventors on the patent application and as an author on the scientific report published by the American Chemical Society. He also worked on projects in the area of steroid chemistry and took graduate courses at night school at both NYU and Brooklyn Polytechnical Institute. He realized that part-time graduate work was a poor way to get a doctoral degree. Djerassi applied to the Wisconsin Alumni Research Foundation for a fellowship. He was awarded the fellowship and CIBA Corporation gave him a small grant.

At the University of Wisconsin, he was influenced by two steroid chemists, William S. Johnson and Alfred L. Wilds. The latter became Djerassi's dissertation adviser and the former a valued friend. His dissertation concerned the transformation of testosterone, the male hormone, into a female hormone, estrogen.

Djerassi received his Ph.D. from Wisconsin in two years and returned to his job at CIBA. CIBA was a generous employer and allowed its scientists to work on their own project for one day each week. Djerassi was able to publish reports on his own research. He hoped to gain a faculty appointment and publications are necessary to fulfill that ambition. In spite of his good work, he was not offered a position by any top-notch academic institution.

So, Djerassi remained at CIBA. By 1949, the therapeutic properties of cortisone were being investigated. Djerassi wanted to work on cortisone synthesis because of his interest in steroid chemistry. However, CIBA's managers decided to conduct their cortisone research at their headquarters in Switzerland. He was disappointed in this development and began to look around for a position that would give him more freedom. He heard of an opening as a codirector of research at Syntex, a small pharmaceutical company in Mexico City.

A little-known firm such as Syntex seemed a foolish move for someone as ambitious as Djerassi. However, they paid for his visit to Mexico City and he could not refuse the offer of an interesting journey. When he arrived at Mexico City, he was met by another expatriate Eastern European, the Hungarian George Rosenkrantz. The two shared more than an Eastern European background. They were both young and highly ambitious.

The Syntex laboratory was well equipped—far better than Djerassi had expected. Perhaps the combination of Rosenkrantz, Djerassi, and the Syntex Company would gain the shining reputation that they all craved. Djerassi accepted the appointment in spite of the skepticism of his former colleagues at the University of Wisconsin.

At that time, Djerassi's friends did not know that Rosenkrantz had furthered the pioneering work of Russell Marker. Not only did Syntex use the yam to supply progesterone to large drug companies, but Rosenkrantz had employed Marker's ideas to produce the male hormone testosterone from the same Mexican yams. Because testosterone was an expensive material, the little company was quite prosperous before Djerassi arrived on the scene. This good cash flow allowed Rosenkrantz and Djerassi to operate a double shift in the laboratory. Each senior researcher had a team of four chemists to conduct the time-consuming work involved in shifting the molecular structure of the diosogenin to that of the hormone. By June of 1951, they succeeded in producing pure cortisone from the diosogenin. This achievement made Syntex a world leader in steroid chemistry —but Djerassi was not finished.

In a few months' time, Djerassi and his team of chemists had synthesized a hormone called norethindrone that stops the production of eggs in human females. This hormone became the foundation for the production of an oral contraceptive. By the early 1960s, Syntex was the major supplier for that rapidly growing market.

By this time, Syntex had become a research partner with a major U.S. drug firm and soon was purchased by financiers based in New York City. The Mexican government attempted to extract as much money as possible from the successful Syntex Company. They imposed high tariffs on the export of the steroids and the raw yams. Then, the government took over yam production in Mexico. With this act, the government had gone too far. U.S. drug companies that produced or marketed steroids found alternative sources for their ingredients. The investors encouraged Djerassi to move to California, take a faculty position at Stanford University in Palo Alto, and continue his research away from Mexican officials.

## Percy L. Julian

In the meantime, other outstanding biochemists were using a different approach to the problem of cortisone production. One such noted scientist was Percy Julian.

Percy Julian was born in the spring of 1899 in Montgomery, Alabama, as the oldest of six children. His father was a railway mail clerk and his grandfather and great-grandfather had been slaves. Both the old men were alive when Percy was growing up and their wit and knowledge inspired him in many ways.

At that time, there were no high schools for African-American children in Montgomery. The nearest secondary school had been established to train teachers for the segregated elementary schools. Percy and his family hoped that this school would provide a good preparation for college. They were disappointed in the curriculum because it offered little instruction in the sciences. Indeed, when Percy was offered a partial scholarship at DePauw University in Greencastle, Indiana, he was not properly prepared for college-level work. He was admitted on probation and remained in that situation for almost three years.

During his entire college career, he supplemented his small stipend by waiting tables at a fraternity house. Some of the time, he worked as a laborer during the day and attended classes in the evenings. Despite these difficulties, Percy loved the school. He encouraged his two brothers and his three sisters to enroll at DePauw. Ultimately, the entire family—mother, father, sisters and brothers—all moved to Greencastle.

After he graduated in 1920, Julian was employed by Fisk University as an instructor in chemistry. While at Fisk, he encouraged many young African Americans to seek careers in science.

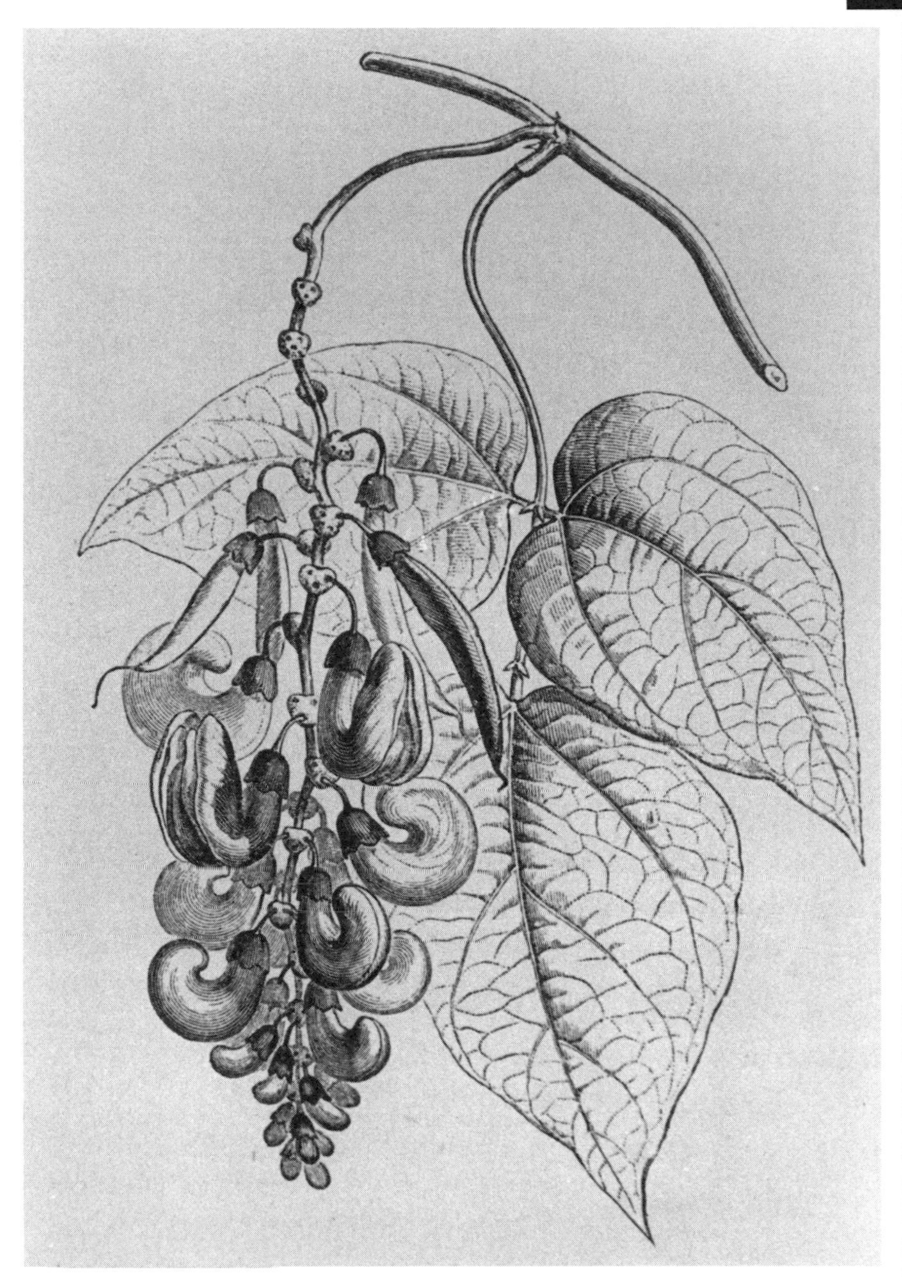

*The poisonous beans of the Calabar vine are the source of eye medicine and were formerly used in tribal ceremonies to determine the guilt or innocence of a person accused of a crime.* (Courtesy of the New York Botanical Garden)

In 1922, Percy Julian was awarded a fellowship to Harvard University and received his master's degree in 1923. For a time, he worked at Harvard as a research assistant to one of the top organic chemists. Following that job, he taught at an all-black school, West Virginia State College.

Another fellowship was soon available to Julian. This was a prestigious Rockefeller grant to study for a doctoral degree in chemistry at the Chemische Institut in Vienna, Austria. After receiving his doctorate, Julian collaborated for several years with an Austrian colleague, Josef Pikl.

Julian spent two years at Howard University in Washington, D.C., after returning from Europe. His next position was at DePauw, his undergraduate college. Julian taught in the chemistry department and conducted research. This was a turning point for him. Research became so vitally important to Julian that he and his friend Pikl spent all their time working on the delicate synthesis of physostigmine. This alkaloid, derived from the Calabar bean, is used in the treatment of glaucoma, an eye disease that can cause blindness.

Julian and Pikl were soon in competition with a team of British scientists. The British team was working on an alternative approach to the synthesis of physostigmine. Julian disputed some of their results. In fact, he stated publicly that the British researchers were wrong. His career was at risk unless he could prove his accusations. To do that, he had to show that he had assembled the correct molecule of physostigmine and the British had produced something else. In February 1935, the two synthetic materials were matched with the natural material. Julian's product was shown to be identical.

The oil of the Calabar bean is also a source of stigmasterol—another natural steroid from which sex hormones can be constructed. Julian succeeded in isolating the steroid from the oil by using a mild acid wash. He wanted to evaluate the effectiveness of the wash by using soybean oil, a more plentiful and less expensive raw material. Julian ordered 5 pounds (2 kg) of soybean oil from the Glidden Company to carry on his research. Shortly, he received his shipment of the oil and a job offer from

*Percy Julian found a way to extract cortisone from soybean oil.* (Courtesy of the Glidden Company and the American Chemical Society)

Glidden. The company bosses had been following Julian's career and had been impressed with his ability—especially his handling of the British competition. He decided to postpone his acid wash tests and accepted Glidden's proposal. The new job led to other research on soybean oil, which later produced both sex hormones and cortisone.

11

# A Modern Plant Hunter

In the mid-1800s, Richard Spruce had demonstrated that the peoples of the Amazon area were highly astute in their identification of medicinal plants. He spent most of his adult life in South America and faced many problems—such as bouts with malaria—to bring back hundreds of plants for study by European scientists. For many years, no one followed in his footsteps. Finally, a young botanist from Harvard took up the challenge.

## Richard Evans Schultes

Dick Schultes was born on January 12, 1915, in Boston, Massachusetts. He grew up in East Boston, a working-class neighborhood across the Chelsea Creek from the center of the city. His mother's father was a master mechanic at a local shipyard. His father's father was a former German military officer who worked as a drayman for a brewery, delivering barrels of beer to local pubs.

As an adolescent, Schultes was not very sociable, but he was driven to excel in his school work. He applied to only one college, Harvard, and was accepted. During his first year at college, he was on a tight budget. His money problems were eased when he received a Cudworth Scholarship, which provided tuition funds and a small stipend.

Schultes continued to work part-time for pocket money. He filed cards and shelved books at the Harvard Botanical Museum Library. The director, Oakes Ames, taught a course on practical botany and Schultes enrolled in the class during his third year of college. Ames was not a good lecturer and attracted few students. In fact, there were only five students in Schultes's class. The small class size meant that Ames could give personal attention to each student. He designed individual research projects that were equal to graduate-level assignments.

By chance, Schultes chose to study peyote, a small spineless cactus native to Mexico and the southwestern United States. Little was known about this plant at the time. Schultes soon discovered that the crown of the cactus—the only part that shows above ground—forms a small buttonlike structure. American Indians chew these "buttons" in some of their religious ceremonies and the cactus buttons cause hallucinatory visions. Schultes was fascinated by the idea that the little plants could have such a dramatic effect on the human mind. He decided to write his undergraduate thesis on peyote.

Ames indicated that Schultes needed to do field research before he could begin to organize his thesis. Therefore, the young man set out on a journey—financed by the independently wealthy Ames—to the American Southwest. In 1936, Schultes visited the Kiowa, one of the Indian tribes who used peyote, and observed the ceremonies where cactus buttons were a part of the ritual.

Peyote has been used in North and South America for untold centuries. In 1500s, when the Spaniards occupied Mexico, Catholic missionaries set out to convert the native population to Christianity. Later, members of the Inquisition—Spanish-organized, Catholic tribunals that disciplined disbelievers by captivity, punishment, or death—came to Mexico. Followers of the old Native American religions were severely punished and the use of peyote was expressly forbidden. However, some of the tribes lived in isolated areas and old ways were preserved, protected from the Spanish priests and soldiers.

Years later, the U.S. government attempted to suppress the use of peyote in the religious ceremonies of the Native American Church—a recognized denomination in the United States. Indeed, anti-peyote laws were passed in nine states. In reaction to these restrictions, the use of peyote by Native Americans spread throughout the West and into Canada.

So it was that in the early summer of 1936, Dick Schultes and another young ethnobotanist, Weston La Barre, a graduate student from Yale, arrived in Oklahoma to explore the use of peyote by local Native Americans. Both young men were working under the direction of Alexander Lesser, a professor of anthropology at Columbia University. Lesser had arranged for Charlie Charcoal, a member of the Kiowa tribe, to join the group. The two students interviewed many tribal elders and eventually were admitted into a peyote ritual. Schultes's visions were limited to a succession of bright colors. When dawn came, he threw up violently but soon recovered. La Barre experienced more dramatic hallucinations. Although Schultes was somewhat disappointed in his experience, he developed a respect for the ceremony and later defended the use of peyote in several legal battles.

His next research subject was another mind-altering plant, the sacred mushroom of the Aztec. Some prominent botanists believed that stories of the mushroom were myths told by imaginative natives since the time of the *conquistadores*—the Spanish soldiers who conquered South America in the 1500s. However, Schultes, while working in the National Herbarium in Washington, D.C., had come across evidence that the stories were true. Dr. Ames was convinced that this student could find additional evidence in tropical Mexico and, again, Ames financed Schultes's research.

South of Mexico City and beyond the end of the railroad line, Schultes located the isolated villages of the Mazatec people. He soon verified the effects of the sacred mushrooms by observing a ceremony that combined religious visions, divination (predicting the future), and healing rituals. Schultes saw that the Mazatec's use of the mushroom closely paralleled the Plains Indians' use of peyote.

Schultes's next research project was conducted deep in the rain forests and concerned pre-Columbian practices of the Inca. He had located old references to a substance called by the Indian name *ololiuqui.* Prominent botanists believed that this plant was a relative of the thorn apple. Schultes thought otherwise. To him, the writings clearly showed that the plant was a climbing vine.

Schultes traveled by mule to the land of the Chinantec deep in the rain forests. In a small village, he visited the local healer and saw that his house was completely overgrown with a vine. Schultes learned that the seeds from this vine constituted the healer's sole medicine and that he sold the seeds to other shamans. The vine fit the Aztec description of *ololiuqui* in every detail. As Schultes had suspected, the historic plant was a vine of the morning glory family.

After his return to Harvard in the fall of 1939, Schultes worked on his doctoral dissertation and recuperated from an infection and the stresses of life in the jungle. He completed his dissertation and began a project sponsored by the Guggenheim Foundation. Schultes traveled back to the Amazon country to study the sources of arrow poisons used by the local peoples.

Word of the Japanese attack on Pearl Harbor in Hawaii reached Schultes while he was staying in the Colombian town of Macoa. The attack caused the United States to enter World War II. Schultes immediately traveled to Bogotá to determine how he might help the war effort. However, the U.S. Embassy in Bogotá was in a confused state. After waiting two months for an assignment, Schultes returned to the Amazon area to continue his research on arrow poisons and other local plant-derived materials.

His progress was good. The young botanist identified a plant product used by the natives to cover wounds and lacerations. Known as "dragon's blood," the resin of the plant forms an antiseptic liquid bandage and the wounds heal with remarkable rapidity. Schultes also investigated a mood-altering substance called *yage*. This material is brewed as a tea from bark scrapings of a thick vine.

After his three-month survey of the recipes for arrow poison, Schultes returned to Bogotá. He had not yet received an official assignment from the U.S. government. Back into the rain forest he went—still under sponsorship of the Guggenheim Foundation—and resumed work on his botanical surveys. This time, his journey took him into the lower reaches of the Amazon, where wild rubber trees had once been exploited for European markets.

World War II had disrupted the sources of many vital raw materials. The great rubber plantations of Malaysia and Indonesia were overrun by the Japanese at the start of the war. The rubber situation became worse as the war progressed. The development of synthetic rubber was incomplete, and the production of war materials required vast amounts of the material. Recycling of worn-out tires and other sources supplied a small fraction of the need. In desperation, officials of the U.S. government decided to try to reopen the wild rubber production of the Amazon basin. They sought to identify the most productive wild trees so that new plantations could be established in Puerto Rico and the countries of Central America. Schultes and other botanists were given the responsibility to identify the best stock and collect seeds for new plantings. At last, Schultes had received an official assignment from the government.

The botanists soon realized that a recurring blight could attack and kill cultivated rubber trees. For this reason, only wild species had been utilized in Brazil during the heyday of the rubber boom. When the trees were cultivated in narrow rows, the blight spread quickly from tree to tree and whole plantations died out in one season. Wild trees were spaced far enough apart so that one diseased tree would not automatically infect the whole crop. Rubber trees in Malaysia were safe because the spores of the blight were too fragile to survive the long trip from the Amazon to East Asia.

For the next four years, Schultes worked on the problem of cultivating rubber trees. Indeed, he continued his research after

*Richard Schultes has been a leading figure in ethnobotany for more than 50 years.* (Courtesy of the Botanical Museum, Harvard University)

World War II had drawn to a close. However, the venture was never truly successful.

When Schultes returned to Harvard in 1953, he became curator of the Orchid Herbarium at the Botanical Museum. His early contributions to the field of botany included studies of peyote, the magic mushroom, the seeds of the morning glory, and the bark of the vine used to brew *yage*. Schultes brought careful scientific analysis to the investigation of mood-altering substances native to Central and South America.

During the 1960s and 1970s, while working with students at the herbarium, Schultes resumed his research on medicinal compounds. He and his students studied the role of coca leaves in the culture of the mountain Indians of South America. Schultes also began work on a compilation of medicinal plants of northwest Amazonia. Researchers estimate that there are as many as 80,000 species of plants in the Amazon jungles. Only a fraction of these species have been screened for various uses.

Schultes, his colleagues, and his students used the techniques of ethnobotany to identify 1,516 species of plants used by the native peoples of South America. Less than half of these plants have been scientifically evaluated. A book, *The Healing Forest*, was published in 1990 and contains descriptions and commentary on all 1,516 species. The volume is coauthored by Schultes and Robert Raffauf.

12

# Branching Out

Officials and scientists in U.S. government service and those in the pharmaceutical industries together with university scholars, freelance explorers, native peoples, and other concerned individuals share an interest in the world's rain forests. Freelance explorers have made dramatic contributions to the array of herbal medicines. The adventures of Nicole Maxwell provide an example of the independent scientist.

By her own account, Maxwell had a diversified educational background that included some medical training and some experience as a professional ballet dancer. She took up serious plant hunting in 1958 when she received a small grant from a major drug company. Maxwell explored the upper Amazon region for two years and made several interesting finds, including an herbal contraceptive. The active ingredient in this compound was thought to be derived from the chopped roots of a plant called *piripiri*.

When Maxwell returned to New York, her sponsors encouraged her to tell radio and television audiences about her exciting experiences as a plant hunter. However, the sponsors showed little interest in her plant discoveries. Possibly, the drug company did not want scientific interest in Maxwell's contraceptive to compete with the interest in their new, and similar, products.

Perhaps early tests failed to show any effectiveness for the material. In fact, later studies suggest that the active ingredient in the contraceptive was provided by a microscopic fungus that infected the roots of the plant. The plant itself may have been totally ineffective.

Nicole Maxwell's experiences with commercial drug companies over a 30-year period reveal the corporations' irregular pattern of interest. Botanical surveys would be launched and then dropped. Sometimes her work included consultations with native healers and sometimes mass screenings of randomly collected plants.

The commercial drug companies were not alone in their interests in natural products. In the early 1950s, the National Cancer Institute began a program of collecting and screening plants for anticancer properties. After the 1960s, the program was aided by botanists and plant specialists from the U.S. Department of Agriculture. About 35,000 plant samples were screened before the program was dropped in 1982. The project found several interesting materials, including taxol—the anticancer drug made from the bark of the yew tree.

In 1986, a revised and modernized program of plant collection and screening was initiated by the National Cancer Institute. The goal was to discover natural plant products that would prove effective against cancers caused by the infections of the AIDS virus. Plant hunters were send to explore the tropical rain forests of Southeast Asia.

The new program included the collaboration of the Cancer Institute with other organizations such as the Field Museum of Natural History and the College of Pharmacy of the University of Illinois—both in Chicago. Also collaborating were herbariums and similar organizations in countries hosting the plant hunters. Funds have been set aside to reimburse governments and provide royalties to tribal groups if local medicinal plants are adoptable into European-style therapies.

In this latest version of plant hunting, organizations such as the Research Triangle Institute of North Carolina and drug companies based in both the United States and Great Britain have been participating in the venture. The areas to be explored now include the South American and African rain forests.

Recently, the U.S. National Science Foundation (NSF) has begun a series of parallel programs that employ partnerships called International Cooperative Biodiversity Groups (ICBGs). Funds are drawn from the National Institutes of Health and the U.S. Agency for International Development as well as the NSF. The partnerships include members from universities, pharmaceutical companies, environmental organizations, and officials from the host countries. The goal of these partnerships is to find new drugs, stop the destruction of the rain forest, and provide employment for the forest people.

Five such groups are working toward this goal. One is in Suriname, a former Dutch colony on the northeastern coast of South America. The Suriname group is split into two teams. One team, from the Missouri Botanical Garden, is using conventional plant-finding techniques to secure medicinal plants for testing. They are collaborating with botanists from the Suriname National Herbarium. The other team is led by an ethnobotanist from Conservation International Suriname. This team has focused their efforts on medicinal plants that have been suggested by shamans. A Surinamese pharmaceutical company prepares extracts from these plants and ships the extracts to the United States for testing and analysis.

The project director of the testing program is David Kingston. He is a natural products chemist from Virginia Polytechnical Institute who worked on the development of taxol. So far, about 30 of the 900 tested plants show some promise. It is too early to determine which of the two methods—conventional plant selection or shaman-directed selection—will be the most fruitful.

The sponsors of these projects have taken care to complete formal agreements with each host country. This will guarantee that governments, villagers, and other participants will receive a fair share in whatever benefits may arise. Great care will be taken if a selected plant proves to have religious significance to any group involved with the project.

In addition, native shamans have been given permission to claim patent rights if plant materials used in their healing rituals are employed in any modern scientific preparation. Pharmaceutical companies are financing a variety of projects to improve the health and well-being of the forest peoples. These projects include bringing electricity into the villages and planning expanded development of forest products in addition to the herbal medicines. Conservation International, a nonprofit philanthropic organization, is taking steps to ensure the continuation of tribal traditions, rituals, healing methods, and history by recording the shamans as they discuss their practices.

In another move to protect native rights, the identity of the plants suggested by the shamans is not revealed unless the active ingredient is proven to be valuable. Then, the parties negotiate an arrangement for the utilization of the compound.

## Doing the Right Thing

The virgin forests of the world, particularly the tropical rain forests, are botanical treasures. They are being thoughtlessly destroyed. Their preservation is important for four reasons. The first two are highly practical. Every year, forest plants absorb tons of carbon dioxide from the atmosphere. This gas, produced by burning fossil fuels such as gasoline and fuel oil, acts as a blanket to hold heat near the surface of the earth. The world's climate could change if the carbon dioxide in the atmosphere continues to increase. By destroying the rain forests, the accumulation of the gas will be accelerated.

The second reason concerns soil erosion. When natural vegetation is cleared from rain forests to provide more pasture and farmland or to harvest timber, the soil becomes vulnerable. The floor of a rain forest is usually a relatively thin layer of soil over a base of bedrock. If the soil is washed away, it is difficult or impossible to restore.

The third and fourth reasons are more theoretical. A healthy forest needs biological diversity—a mix of many different plant and animal species. In contrast, much of modern agriculture uses a so-called monoculture where only one species of plant—such as wheat—is cultivated over a large territory. If a plant disease hits one section of the territory, the disease can spread through the entire area and kill every plant. If the diversity of rain forests were reduced to such a monoculture, they would be at risk of devastation by disease or other natural disasters.

The fourth reason will affect the future of civilization. The reduction of plant species by deforestation decreases the number of possible medicinal plants and other natural products. When our petroleum reserves run out some time in the future, natural plant products will be needed to take their place. Both gasahol and lubricating oils can be made from plant materials. In addition, the destruction of some species means that domesticated plants may lose closely related wild plants. The wild species often have a resistance to disease that domestic varieties have lost. At some time, a renewed breeding with wild species will probably be needed to restore the vitality of domestic crop plants.

## Neem

Discoveries based on folk medicine continue to appear. The neem tree, which grows wild in India, has recently gained prominence in the industrialized world as a possible medicinal plant. Today, the neem tree is being adopted as a new cultivated crop in tropical areas such as Africa and the Caribbean islands.

For centuries, villagers in India used the leaves and seeds of this tree for health-care purposes. They use neem twigs, for example, to brush their teeth. The underbark of these twigs contains antibacterial chemicals and the twig toothbrushes prevent gum disease.

Villagers have long used neem leaves as insect repellents. Leaves and small branches of neem are added to grain stores.

Tests by the U.S. Department of Agriculture and several other countries confirm that many insect predators are repelled by sprays made from water or alcohol containing an extract from neem leaves. Insects that are exposed to neem sprays are not killed immediately, but their biological functions are severely impaired. Some insect species are made sterile by such treatments.

Oil from neem seeds appears to be an effective treatment for skin problems. Chemicals in the oil are toxic to both bacteria and viruses. The drug-regulating agency of the Indian government has recently approved the sale of processed neem oil as a contraceptive cream.

The history of neem is similar to that of the plant *Rauwolfia* and the chemical reserpine. Neem was long used by Indian healers to make traditional medicine, but the plant was ignored in industrialized countries. After Indian scientists conducted the initial testing and development work, other countries became interested in the health-care properties of the tree. If neem fulfills even a fraction of the claims made by present-day enthusiasts, its acceptance could solve many medical and economic problems —particularly in tropical, nonindustrialized countries.

# Glossary

**alkaloid** A family of carbon-based molecules that always include one or more nitrogen atoms. The typical alkaloid has a bitter taste. Caffeine and nicotine are common examples.

**amino acid** A carbon-based molecule that contains nitrogen and is the building block for the assembly of proteins.

**antibiotic** A compound that kills or stops the reproduction of microbes.

**archaeology** The study of the material remains of ancient societies.

**artifact** An object produced by human craft.

**Atabrine** The trade name for a synthetic compound similar to natural quinine.

**black bile** Traditionally, the source of melancholy. Organically, the fluid produced by the gallbladder as an aid to the digestion of fats.

**botulism** An often fatal type of food poisoning caused by microbes.

**cholera** A severe and highly contagious infection of the intestinal system that is often fatal.

**endocrine gland** Organs that produce chemicals that are secreted directly into the bloodstream. The chemicals then control various bodily processes.

**enzyme** Protein-type molecules that direct the chemical reactions in biological cells.

**epileptic seizure** The acute stage of epilepsy characterized by jerky, convulsive movements and unconsciousness.

**gasahol** A mixture of about 90 percent gasoline and 10 percent ethyl alcohol used as fuel for automobiles.

**glaucoma** A condition of impaired vision due to hardening of the eyeball from internal pressure.

**hallucinatory visions** False perceptions that are often very dramatic but convincingly real.

**harbinger** An advance warning; an indicator of events to come.

**herbal** A document describing useful plants and often containing recipes for their use.

**hormone** A chemical that controls, guides, or activates a bodily process.

**insulin** The hormone, produced by glands in the pancreas, that controls the use of sugar by body cells.

**interlocutor** A person who acts as a bridge or interpreter between two other persons; a translator.

**Jesuits** Members of a Roman Catholic order founded in 1534 noted for their discipline and scholarship.

**lethality** The degree or extent of the ability to cause death.

**medical physiologist** A biological scientist who specializes in studies of the bodily organs susceptible to disease.

**mid-latitudes** The areas of the globe that lie between the Arctic Circle and the Tropic of Cancer in the Northern Hemisphere and between the Tropic of Capricorn and the Antarctic Circle in the Southern Hemisphere.

**molecular biology** The study of biological processes at the molecular level with particular emphasis on the assembly of enzymes and other proteins.

**moorland** A broad expanse of open land; often poorly drained and boggy.

**morphine** An addictive alkaloid extracted from opium; used in medicine as a pain killer and sedative.

**palliative** A compound that can reduce the symptoms of a disease but not cure it.

**pharmacology** The study of medicinal materials; particularly their chemical composition, their application and their effects.

**phlegm** Mucus.

**placebo** An inactive substance administered as a comparison check in tests of the effectiveness of a drug.

**proprietary drug** A compound that has its name protected by a registered trademark and that can be dispensed without a doctor's prescription.

**protein** A large, complicated carbon-based molecule with attached nitrogen that is a building block for all animal tissue.

**proximity** Closeness.

**puffer fish** A member of the family of fish that are able to swell in size when disturbed.

**sauna** A steam bath originated in Finland that is generated by splashing water on heated rocks.

**steroid** A carbon-based molecule that contains an alcohol attachment and is fat soluble. Steroids provide the basic molecule for the assembly of many hormones and vitamins.

**sushi** Various dishes having raw fish as a primary ingredient.

**therapeutic claims** The promises of medical effectiveness; often inflated for proprietary drugs.

**tranquilizer** A medical compound that reduces anxiety, calms, and relaxes the patient.

**tuberculosis** A disease that can infect both humans and other animals. The symptoms include lesions of the lining of the lungs.

**vedic medicine** The medical materials and procedures developed within the Hindu culture of India.

**yellow bile** Choler, a product of the gallbladder needed for the digestion of fats.

**yellow fever** An infectious disease caused by a virus and characterized by jaundice or the development of a yellowish body color.

# Further Reading

Barrett, S., and Gilda Knight. *The Health Robbers*. Philadelphia, Pa.: G.F. Stickley Co., 1993. Relates the ways in which dealers in patent medicines and nostrums have taken advantage of sick people who are desperate for relief.

Crellin, John K., and Jane Philpott. *Herbal Medicine Past and Present*. Durham, N.C.: Duke University Press, 1989. This is a source book for studies of the people of rural America and their shared attitudes about health care.

Davis, Wade. *One River*. New York: Simon & Schuster, 1996. The author weaves back and forth between the story of his recent travels with the ethnobotanist Tim Plowman in the upper Amazon region and the key events in the South American expeditions carried out in the 1930s and 1940s by Richard Evans Schultes.

Friedman, David. *Focus on Drugs and the Brain*. Frederick, Md.: Twenty-First Century Books, 1990. Takes a candid approach to the effects of various drugs. A readable and interesting discussion for young people.

Huntley, Beth. *Amazon Adventure*. Milwaukee, Wisc.: Gareth Stevens, 1989. Colorfully illustrated virtual trip down the Amazon with attention to wildlife, plants, and native peoples.

Joyce, Christopher. *Earthly Goods*. Boston, Mass.: Little Brown & Co., 1994. Covers the contributions of North and South American Indians to the assembly of useful medicines—with particular emphasis on curare and quinine.

Lerner, Carol. *Moonseed and Mistletoe: A Book of Poisonous Wild Plants*. New York: Morrow, 1988. Recommended as a reference source, it focuses on plants of North America.

Maxwell, Nicole. *Witch Doctor's Apprentice*. New York: Citadel Press, 1990. An autobiographical narrative that relates the author's adventures and dealings with other explorers, missionaries, and the native healers.

Van Hagen, Victor W. *South America Called Them*. New York: A.A. Knopf, 1945. The adventures of La Condamine, Humboldt, and Spruce plus those of Charles Darwin.

# Index

Page numbers in *italics* indicate illustrations.

615
Kid